Quantitative Methods and Applications in GIS

Fahui Wang

Taylor & Francis
Taylor & Francis Group
Boca Raton   London   New York

A CRC title, part of the Taylor & Francis imprint, a member of the
Taylor & Francis Group, the academic division of T&F Informa plc.

Published in 2006 by
CRC Press
Taylor & Francis Group
6000 Broken Sound Parkway NW, Suite 300
Boca Raton, FL 33487-2742

International Standard Book Number-10: 0-8493-2795-4 (Hardcover)
International Standard Book Number-13: 978-0-8493-2795-7 (Hardcover)
Library of Congress Card Number 2006040460

Library of Congress Cataloging-in-Publication Data

Wang, Fahui, 1967-
    Quantitative methods and applications in GIS / Fahui Wang.
        p. cm.
    ISBN 0-8493-2795-4
    1. Geographic information systems--Mathematical models. I. Title.

G70.212W36 2006
910.285--dc22   2006040460

Taylor & Francis Group
is the Academic Division of Informa plc.

Visit the Taylor & Francis Web site at
http://www.taylorandfrancis.com

and the CRC Press Web site at
http://www.crcpress.com

Dedication

In loving memory of Katherine Z. Wang
To Lei and our three J's (Jenny, Joshua, and Jacqueline)

Foreword

This splendid book argues that to do good social science that is policy relevant, quantitative methods are essential and such methods, and the theory behind their practice, must be spatial. Accordingly Fahui Wang sets out to show how relevant applications at the level of cities and regions must be fashioned using the methods of quantitative geography which are currently best expressed in GIS (geographic information systems) and GI science. What is nice about his approach is that he grounds all the methods that he introduces in practical applications that are supported by the data files used in the examples, presented in such a way that readers at both the beginning and more advanced levels can design and explore their own simulations.

In the last decade, GIS has come of age and its synthesis and co-development with spatial analysis and quantitative geography is generating an edifice that has come to be known as GI science. This science is not simply method- or technique-driven, for it relates strongly to geographical theory, whether it be from the social or the physical domain or both. This book mainly deals with social (and economic) applications but the methods used are not restricted to the social world. Far from it. Spatial analytic method is being developed in many fields where geographical space of various kinds — topological, Euclidean, in any dimension, and so on — is invoked. Moreover, several of the methods introduced here for social applications emerged originally from the physical and natural sciences, in the geophysical, medical, and ecological realms, for example. A synthesis is in fact being forged with computational science where the focus here is on computational social science as an essential apparatus in the development of social understanding and social policy.

There are several key themes exploited in this book which serve to define the spatial domain. In particular, the idea of distance, proximity and accessibility are central to ways of defining concentration and dispersion in space through clustering, density, homogeneity, and hinterland. These serve to illustrate the form and function of urban and regional systems at a variety of scales and the techniques developed around these foci all enable the physical and social morphology of cities in their regions to be measured and analysed consistently. This is GI science in the making, and throughout this book the author is at pains to emphasise how functions and forms, which at first sight might appear disparate, link together in more generic systems and models. The applications that are developed here range over several urban sectors and scales from health care and crime to transportation and retailing. The focus, too, is not simply on measurement and understanding, for all the examples are set within a policy context which presupposes problems to be solved. Indeed toward the end of the book, there are applications dealing with formal optimisation that generate specific and unique solutions to various spatial problems, particularly in transportation.

In fact one of the key concerns in this book is to identify how key policy problems, whether they are in terms of finding the best location for a shopping center or identifying a critical cluster of diseases, are articulated using spatial analysis. These kinds of problem are increasingly amenable to such quantitative analysis largely because of better and more widely available data sources at ever finer scales, and because we now have technologies that are able to rapidly synthesize and visualize the meaning of different patterns implicit in such spatial data. This is what GIS has brought to this science and it is no accident that quantitative analysis in the social sciences is now being quite heavily informed by the spatial perspective. It is hard for example to now undertake a study of patterns of disease and its mitigation through better health care without using spatial data. Moreover in a world where resources are limited in the face of better methods for identifying problems and where the world is becoming ever more complex because of new technologies and increasing personal opportunities, such spatial analysis becomes essential. This is another motivating theme in this book which serves to impress on the reader how important it is to develop sound analysis in space for problems that traditionally have hardly merited any kind of spatial analysis. Crime is an excellent example, and Fahui Wang shows quite convincingly how one can make good progress in using techniques developed originally for problems of clustering in soil science and geology, first in the identification of clusters of diseases and then in the all important analysis of crime hot spots. This immediately generates interest in policy questions. What the author is able to do most effectively here is to illustrate the ways in which quite routine methods can be adapted to identify important problems which have wide policy relevance.

At various points in this book, more comprehensive models are introduced. In fact, models of retailing and population density combined with accessibility analysis and operationalised through spatial interaction, emerge as comprehensive land use–transport models toward the end of the book. This is a nice feature because it suggests that GI science is a much wider edifice than merely a tool box of techniques in that it is increasingly extending to systems of more general concern and import. The methods and applications here link this work to ideas about the intrinsic nature of such systems and although most of the treatment is focused on spatial analysis in a policy-relevant context, there are glimpses of a wider complexity in city and regional systems that GI science is beginning to respond to.

Michael Batty
Centre for Advanced Spatial Analysis
University College, London

Preface

One of the most important advancements in recent social science research (including applied social sciences and public policy) has been the application of quantitative or computational methods in studying the complex human or social systems. Research centers in *computational social sciences* have flourished on major university campuses. Among others, the University of Chicago, University of Washington, UCLA, and George Mason University have all established such a center recently to promote the multidisciplinary research related to social issues. Many conferences have also been organized around this theme. *Geographic Information Systems* (GIS) has played an important role in this movement because of its capability of integrating and analyzing various datasets, in particular spatial data. The Center for Spatially Integrated Social Science at UC–Santa Barbara, funded by the National Science Foundation, has been an important force in promoting the usage of GIS technologies in various social sciences. The growth of GIS has made it increasingly known as geographic information science (GISc), which covers broader issues such as spatial data quality and uncertainty, design and development of spatial data structure, social and legal issues related to GIS, and many others. On October 20, 2005, Harvard University announced the establishment of a new Center for Geographic Analysis after elimination of the geography program over half a century ago. What has brought geography back to Harvard? It is spatial analysis and geographic information systems (see "Report to the Provost on Spatial Analysis at Harvard University" by the Provost's Committee on Spatial Analysis, Harvard University, 2003).

Many of today's students in geography and other social science-related fields (e.g., sociology, anthropology, business, city and regional planning, public administration) all share the same excitement surrounding GIS. But their interest in GIS may fade away quickly if the GIS usage is limited to managing spatial data and mapping. In the meantime, a significant number of students complain that courses on statistics, quantitative methods, and spatial analysis are too dry and feel irrelevant to their interests. Over the years of teaching GIS, spatial analysis, and quantitative methods, I have learned the benefits of blending them together and practicing them in case studies using real-world data. Students can sharpen their GIS skills by applying some GIS techniques to detecting hot spots of crime, or gain better understanding of the classic urban land use theory by examining spatial patterns in a GIS environment. When students realize that they can use some of the computational methods and GIS techniques to solve a real-world problem in their own field, they become better motivated in class. In other words, *technical skills* in GIS or quantitative methods are learned in the context of addressing *subject issues*. Both are important for today's competitive job market.

This book is the result of my efforts of *integrating GIS and quantitative (computational) methods*, demonstrated in various *applications in social sciences*. The applications are chosen with three objectives in mind. The first is to demonstrate the *diversity of issues* where GIS can be used to enhance the studies related to social issues and public policy. Applications range from typical themes in urban and regional analysis (e.g., regional growth patterns, trade area analysis) to issues related to crime and health analyses. The second is to illustrate *various computational methods*. Some may be cumbersome or difficult to implement without GIS, and others may be integrated into GIS and become highly automated. The third objective is to cover common tasks (e.g., distance and travel time estimation, spatial smoothing and interpolation, accessibility measures) and major issues (e.g., modifiable areal unit problem, rate estimate of rare events in small population, spatial autocorrelation) that are encountered in *spatial analysis*.

One important feature of this book is that each chapter is *tasks driven*. Methods can be better learned in the context of solving real-world problems. Although each method is illustrated in a special case of application, it can be used to analyze different issues. Each chapter has one subject theme and introduces the method (or a group of related methods) most relevant to the theme. For example, linear programming is introduced to solve the problem of wasteful commuting; systems of linear equations are analyzed to predict urban land use patterns; spatial regression is used to examine the relationship between job access and homicide patterns; and cluster analysis is conducted in examining cancer patterns.

Another important feature of this book is the emphasis on *implementation of methods*. All GIS-related tasks are illustrated in the ArcGIS platform, and most statistical analyses (including linear programming) are conducted by SAS. In other words, one may only need access to ArcGIS and SAS in order to replicate the work discussed in the book and conduct similar research. ArcGIS and SAS are chosen because they are the leading software for GIS and statistical analysis, respectively. Some specific tasks, such as spatial clustering and spatial regression, use free software that can be downloaded from the Internet. Most data used in the case studies are *public accessible* (i.e., free online). Instructors and advanced readers may use the data sources and techniques discussed in the book to design their class projects or craft their own research projects. A CD containing all data and sample computer programs is enclosed (see the List of Data Files).

This book intends to mainly serve students in *geography, urban and regional planning*, and related fields. It can be used in courses such as (1) spatial analysis, (2) location analysis, (3) applications of GIS in business and social science, and (4) quantitative methods in geography. The book can also be useful for researchers outside of geography and planning but using GIS and spatial analysis in their studies. Some in *urban economics* may find the studies on urban structures and wasteful commuting relevant, and others in *business* may think the chapters on trade area analysis and accessibility measures useful. The case study on crime patterns may interest *criminologists*, and the one on cancer cluster analysis may find an audience among *epidemiologists*.

The book has 11 chapters. Part I includes the first three chapters, covering some generic issues such as an overview of data management in GIS and basic spatial analysis tools (Chapter 1), distance and travel time measurement (Chapter 2), and spatial smoothing and interpolation (Chapter 3). Part II includes Chapters 4 through 7, covering some basic quantitative methods that require little or no programming skills: trade area analysis (Chapter 4), accessibility measures (Chapter 5), function fittings (Chapter 6), and factor analysis (Chapter 7). Part III includes Chapters 8 through 11, covering more advanced topics: rate analysis in small populations (Chapter 8), spatial cluster and regression (Chapter 9), linear programming (Chapter 10), and solving a system of linear equations (Chapter 11). Parts I and II may serve an upper-level undergraduate course. Part III may be used for a graduate course. It is assumed that readers have some basic GIS and statistical knowledge equivalent to one introductory GIS course and one elementary statistical class.

Each chapter focuses on one computational method except for the first chapter. In general, a chapter (1) begins with an introduction to the method, (2) discusses a theme to which the method is applied, and (3) uses a case study to implement the method using GIS. Some important issues, if not directly relevant to the main theme of a chapter, are illustrated in appendixes. Many important tasks are repeated in different projects to reinforce the learning experience (see the Quick Reference for Spatial Analysis Tasks and Quantitative Methods).

Undertaking the task of writing a book takes courage, perhaps more naivety in my case. I have found myself more often than not falling behind various deadlines and being absent from many family hours. My wife has spared me from much of the housekeeping work. I often hear my kids whispering to each other: "Be quiet! Daddy is working on his book." So foremost, I thank my family for their support and encouragement.

My interest in quantitative methods has very much been influenced by my doctoral advisor, Jean-Michel Guldmann, in the Department of City and Regional Planning of the Ohio State University. I learned linear programming and solving a system of linear equations in his courses on static and dynamic programming. I also benefited a great deal from my acquaintance of Donald Haurin in the Department of Economics of the Ohio State University. The topics on urban and regional density patterns and wasteful commuting can be traced back to his teaching of urban economics. Philip Viton, also in the Department of City and Regional Planning of the Ohio State University, taught me much of the econometrics. I only wish I could have been a better student then.

I am grateful to Northern Illinois University for granting me a sabbatical leave in the fall of 2004, when I began writing the book. I am also indebted to my colleagues Richard Greene, Andrew Krmenec, and Wei Luo for many intellectual conversations and helpful comments. I appreciate the help from Lan Mu at Department of Geography, University of Illinois–Urbana-Champaign, for developing the scale-space cluster tool in Chapter 8. Leonard Walther at the Geography Department of Northern Illinois University helped me design, improve, and polish some of the graphics. Holly Liu at the Public Works Department of City of Geneva, Illinois, digitized the hypothetical city used in Chapter 11. Her generous help and professional work ensured the quality of case study 11. I thank Michael Batty for graciously writing the Foreword on a short notice.

Finally, I would like to thank the editorial team at Taylor & Francis: acquisition editors Randi Cohen and Taisuke Soda, project coordinator Theresa Delforn, project editor Khrysti Nazzaro, and many others including typesetters, proofreaders, cartographers, and computer specialists. Thank you all for guiding me through the whole process.

The case studies in the book have been tested multiple times by me, and also by students who took my Location Analysis, Urban Geography, and Transportation Geography classes at the Northern Illinois University. Most recently during the proof-review stage, I used some of the projects in the workshops on "GIS-Based Quantitative Methods and Applications in Socioeconomic Planning Sciences" in Tsinghua University and China Northeast Normal University, both in China, and received many valuable and positive feedbacks. Many errors may remain. I welcome comments from researchers, teachers and students who use the book. I hope for a chance to revise the book and have a new version in the near future.

The Author

Fahui Wang is associate professor at the Department of Geography, Northern Illinois University. He earned his B.S. in geography from Peking University, China, and his M.A. in economics and Ph.D. in city and regional planning, both from the Ohio State University. His research has been funded by the National Institute of Justice, National Cancer Institute, U.S. Department of Health and Human Services, and U.S. Department of Housing and Urban Development. He has published over 30 refereed articles. In addition to this book, he is also the editor of the book *Geographic Information Systems and Crime Analysis*, published in 2005 by IDEA Group Publishing.

List of Figures

CHAPTER 1

Figure 1.1   Dialog windows for projecting a spatial dataset 6
Figure 1.2   Dialog window for updating area in shapefile 6
Figure 1.3   Attribute join in ArcGIS 7
Figure 1.4   Population density pattern in Cuyahoga County, Ohio, 2000........9
Figure 1.5   Dialog window for spatial join 13
Figure 1.6   Rook contiguity vs. queen contiguity 15
Figure 1.7   Workflow for defining queen contiguity......................................16

CHAPTER 2

Figure 2.1   An example for the label-setting algorithm................................22
Figure 2.2   Three provinces, four major cities, and railroads in
             northeast China...25
Figure 2.3   Three segments in measuring travel distance 26
Figure 2.4   Table joins in computing travel distances 30
Figure A2.1 A valued-graph example 32

CHAPTER 3

Figure 3.1   The FCA method for spatial smoothing.......................................36
Figure 3.2   Kernel estimation 37
Figure 3.3   Tai and non-Tai place-names in Qinzhou 39
Figure 3.4   Tai place-name ratios in Qinzhou by the FCA method 40
Figure 3.5   Kernel density of Tai place-names in Qinzhou 41
Figure 3.6   Interpolated Tai place-name ratios in Qinzhou by
             trend surface analysis 46
Figure 3.7   Interpolated Tai place-name ratios in Qinzhou by the
             IDW method..47
Figure 3.8   Areal weighting interpolation from census tracts to
             school districts..50

CHAPTER 4

Figure 4.1   Constructing Thiessen polygons for five points 58
Figure 4.2   Breaking point by Reilly's law between two stores.....................58

Figure 4.3   Proximal areas for the Cubs and White Sox 64
Figure 4.4   Probabilities for choosing the Cubs by Huff model 67
Figure 4.5   Proximal areas for four major cities in northeast China.............70
Figure 4.6   Hinterlands for four major cities in northeast China by
              Huff model 72

CHAPTER 5

Figure 5.1   An earlier version of the FCA method...80
Figure 5.2   The 2SFCA method 82
Figure 5.3   Procedures in implementing the 2SFCA method........................87
Figure 5.4   Accessibility to primary care physician in Chicago region by
              2SFCA (20 mile)..88
Figure 5.5   Accessibility to primary care physician in Chicago region by
              2SFCA (30 minute)..90
Figure 5.6   Accessibility to primary care physician in Chicago region by
              gravity-based method ($\beta = 1$) 92
Figure 5.7   Comparison of accessibility scores by the 2SFCA and
              gravity-based methods...94

CHAPTER 6

Figure 6.1   Regional growth patterns by the density function approach......100
Figure 6.2   Excel dialog window for regression 103
Figure 6.3   Excel dialog window for adding trend lines 104
Figure 6.4   Illustrations of polycentric assumptions 108
Figure 6.5   Population density surface and job centers in Chicago,
              six-county region...112
Figure 6.6   Density vs. distance exponential trend line (census tracts)..........114
Figure 6.7   Density vs. distance exponential trend line (survey townships)...118

CHAPTER 7

Figure 7.1   Scree graph for principal components analysis..........................130
Figure 7.2   Data processing steps in principal components factor analysis ...131
Figure 7.3   Dendrogram for the clustering analysis example 132
Figure 7.4   Conceptual model for urban mosaic..136
Figure 7.5   Study area for Beijing's social area analysis.............................137
Figure 7.6   Spatial patterns of factor scores..141
Figure 7.7   Social areas in Beijing 142

CHAPTER 8

Figure 8.1   The ISD method 151
Figure 8.2   An example for assigning spatial-order values to polygons 152
Figure 8.3   An example of clustering based on the scale-space theory 154
Figure 8.4   Census tracts with small populations in Chicago 1990 159
Figure 8.5   Dialog window for the scale-space clustering tool 160
Figure 8.6   A sample area for illustrating the clustering process 161
Figure 8.7   First-round clusters by the scale-space clustering method 162

CHAPTER 9

Figure 9.1   SaTScan dialog for point-based spatial cluster analysis 171
Figure 9.2   Spatial clusters of Tai place-names in southern China 171
Figure 9.3   Colorectal cancer rates in Illinois counties, 1996–2000 176
Figure 9.4   ArcGIS dialog for computing Getis–Ord general G 177
Figure 9.5   Colorectal cancer clusters based on local Moran 179
Figure 9.6   Colorectal cancer hot spots and cold spots based on $Gi*$ 180
Figure 9.7   GeoDa dialog for defining spatial weights 183
Figure 9.8   GeoDa dialog for spatial regression 184

CHAPTER 10

Figure 10.1   Columbus MSA and the study area 195
Figure 10.2   Input and output files in the polygon-based
               location-allocation analysis 205
Figure 10.3   Clinic locations and service areas by polygon-based analysis 206
Figure 10.4   Input and output files in the network-based
               location-allocation analysis 209
Figure 10.5   Clinic locations and service areas by network-based analysis 210
Figure 10.6   Highways in Cuyahoga, Ohio 211

CHAPTER 11

Figure 11.1   Interaction between population and employment distributions
               in a city 222
Figure 11.2   A simple city in the illustrative example 224
Figure 11.3   Spatial structure of a hypothetical city 226
Figure 11.4   Population distributions in various scenarios 228
Figure 11.5   Service employment distributions in various scenarios 229

List of Tables

CHAPTER 1

Table 1.1   Types of Relationships in Combining Tables 4
Table 1.2   Types of Spatial Joins in ArcGIS 11
Table 1.3   Comparison of Spatial Query, Spatial Join, and Map Overlay 12

CHAPTER 2

Table 2.1   Solution to the Shortest-Route Problem 23

CHAPTER 3

Table 3.1   FCA Spatial Smoothing by Different Window Sizes 41

CHAPTER 4

Table 4.1   Fan Bases for Cubs and White Sox by Trade Area Analysis 65
Table 4.2   Four Major Cities and Hinterlands in Northeast China.................69

CHAPTER 5

Table 5.1   Travel Speed Estimations in the Chicago Region 89
Table 5.2   Comparison of Accessibility Measures 91

CHAPTER 6

Table 6.1   Linear Regressions for a Monocentric City 104
Table 6.2   Polycentric Assumptions and Corresponding Functions 109
Table 6.3   Regressions Based on Monocentric Functions
            (1837 Census Tracts)...114
Table 6.4   Regressions Based on Polycentric Assumptions 1 and 2
            (1837 Census Tracts) 116
Table 6.5   Regressions Based on Monocentric Functions
            (115 Survey Townships) 118

CHAPTER 7

Table 7.1   Idealized Factor Loadings in Social Area Analysis 135
Table 7.2   Basic Statistics for Socioeconomic Variables in Beijing
            ($n = 107$) 138
Table 7.3   Eigenvalues from Principal Components Analysis 139
Table 7.4   Factor Loadings in Social Area Analysis....................................139
Table 7.5   Characteristics of Social Areas (Clusters)...................................142
Table 7.6   Zones and Sectors Coded by Dummy Variables 143
Table 7.7   Regressions for Testing Zonal vs. Sectoral Structures
            ($n = 107$) 144

CHAPTER 8

Table 8.1   Approaches to Analysis of Rates of Rare Events in
            Small Population...151
Table 8.2   Rotated Factor Patterns of Socioeconomic Variables in
            Chicago 1990 157
Table 8.3   OLS Regression Results from Analysis of Homicide in
            Chicago 1990 160

CHAPTER 9

Table 9.1   Cancer Incident Rates (per 100,000) in Illinois Counties,
            1986–2000...175
Table 9.2   Global Clustering Indexes for County-Level Cancer
            Incidence Rates..178
Table 9.3   OLS and Spatial Regressions of Homicide Rates in Chicago
            ($n = 845$ Census Tracts) 185
Table 9.4   OLS and Spatial Regressions of Homicide Rates in Chicago
            ($n = 77$ Community Areas)...186

CHAPTER 10

Table 10.1 Location-Allocation Models..202
Table 10.2 Location-Allocation Analysis Results (Polygon Based vs.
            Network Based) 207

CHAPTER 11

Table 11.1 Simulated Population and Service Employment Distributions in
            Various Scenarios 228

List of Data Files

Data are organized under various *study areas*, and one folder may contain data used in multiple case studies. Files under various folders may share the same file names, so it is recommended that you organize projects using the same study area under one folder. All shapefiles are in the zip format (the zip file names are provided in parentheses if they use a different name).

1. The folder `Cleveland` contains data for case studies 1A, 1B, 3C, and 10B:
   - Coverage interchange files: `clevbnd.e00`, `cuyatrt.e00`
   - Shapefiles: `tgr39035trt00` (`trt0039035.zip`), `tgr39035uni` (`uni39035.zip`), `tgr39035lka` (`lkA39035.zip`), `cuyautm`, `cuya_pt`, `clevspa2k`
   - dBase file: `tgr39000sf1trt.dbf`
   - Text files: `Queen_Cont.aml`, `Cuya_hosp.csv`
2. The folder `ChinaNE` contains data for case studies 2 and 4B:
   - Coverage interchange files: `cntyne.e00`, `city4.e00`, `railne.e00`
   - dBase file: `dist.dbf`
3. The folder `ChinaQZ` contains data for case studies 3A, 3B, and 9A:
   - Coverage interchange file: `qztai.e00`
   - Shapefile: `qzcnty`
4. The folder `Chicago` contains data for case studies 4A, 5, 6, 8, and 9C:
   - Coverage interchange files: `chitrt.e00`, `citytrt.e00`, `citycom.e00`
   - Shapefiles: `tgr17031lka` (`lkA17031.zip`), `chitrtcent`, `chizipcent`, `polycent15`, `county6`, `county10`, `twnshp`
   - Text files: `cubsoxaddr.csv`, `monocent.sas`, `polycent.sas`, `cityattr.txt`
   - Program file: `ScaleSpace.dll`
5. The folder `Beijing` contains data for case study 7:
   - Shapefile: `bjsa`
   - Text files: `bjattr.csv`, `FA_Clust.sas`, `BJreg.sas`
6. The folder `Illinois` contains data for case study 9B:
   - Coverage interchange file: `ilcnty.e00`
7. The folder `Columbus` contains data for case study 10A:
   - Coverage interchange files: `urbtazpt.e00`, `road.e00`
   - Text files: `rdtime.aml`, `urbtaz.txt`, `odtime.txt`, `LP.sas`
8. The folder `SimuCity` contains data for case study 11:
   - Coverage interchange files: `tract.e00`, `road.e00`, `trtpt.e00`, `cbd.e00`
   - Text files: `odtime.prn`, `odtime1.prn`, `rdtime.aml`, `SimuCity.FOR`

Quick Reference for Spatial Analysis Tasks and Quantitative Methods

Task[a]	Section First Introduced	Section(s) Repeated
Updating areas for shapefile	Section 1.2	Section 3.6.2, Section 3.6.2, Section 6.5.3
Generating polygon centroids	Section 1.4.1	Section 2.3.1, Section 4.3.1, and others
Computing Euclidean distances	Section 2.3.1	Section 3.2.1, Section 4.3.2, Section 5.4.1, and others
Computing network distances	Section 2.3.2	
Computing travel time	Section 2.3.3	Section 5.4.1, Section 10.2.2, Section 11.3.1
Spatial smoothing by floating catchment area (FCA) method	Section 3.2.1	
Kernel estimation	Section 3.2.2	
Trend surface analysis	Section 3.4.1	
Logistic trend surface analysis	Section 3.4.1	
Spatial interpolation by inverse distance weighted (IDW), thin-plate splines, or Kriging	Section 3.4.2	Section 4.3.2, Section 6.5.1
Areal weighting interpolator	Section 3.6.2	Section 6.5.3
Address matching (geocoding)	Section 4.3.1	Section 10.4.1
Defining proximal areas based on Euclidean distance	Section 4.3.1	Section 6.5.2
Defining proximal areas based on network distance	Section 4.4.1	
Defining trade areas by Huff model	Section 4.3.2	Section 4.4.2
Generating weighted centroids	Section 5.4.1	
Measuring accessibility by 2SFCA or gravity model	Section 5.4.1	
Linear regression in Excel or SAS	Section 6.5.1	Section 7.4, Section 8.4
Function (including nonlinear) fittings in Excel	Section 6.5.1	
Nonlinear or weighed regressions in SAS	Section 6.5.1	

[a]Tasks are implemented in ArcGIS unless otherwise specified.

Task	Section First Introduced	Section(s) Repeated
Principal components and factor analysis in SAS	Section 7.4	Section 8.4
Cluster analysis in SAS	Section 7.4	
Computing weighted averages	Section 8.4	Section 9.6.2
Scale-space melting (regionalization)	Section 8.4	
Point-based spatial cluster analysis in SaTScan	Section 9.2	
Area-based spatial cluster analysis	Section 9.4	
Spatial regression in GeoDa	Section 9.6.1	Section 9.6.2
Linear programming in SAS	Section 10.2.3	
Polygon-based or network-based location-allocation problems	Section 10.4.1, Section 10.4.2	
Solving a system of linear equations in FORTRAN	Section 11.3.2	Section 11.3.3 and others

Contents

PART I GIS and Basic Spatial Analysis Tasks

Chapter 1 Getting Started with ArcGIS:
Data Management and Basic Spatial Analysis Tools...........................1

1.1   Spatial and Attribute Data Management in ArcGIS.....................................1
    1.1.1   Map Projections and Spatial Data Models..................................2
    1.1.2   Attribute Data Management and Attribute Join 3
1.2   Case Study 1A: Mapping the Population Density Pattern in
Cuyahoga County, Ohio..4
1.3   Spatial Analysis Tools in ArcGIS: Queries, Spatial Joins, and
Map Overlays...8
1.4   Case Study 1B: Extracting Census Tracts in the City of Cleveland and
Analyzing Polygon Adjacency..12
    1.4.1   Part 1: Extracting Census Tracts in Cleveland............................12
    1.4.2   Part 2: Identifying Contiguous Polygons 14
1.5   Summary 15
Appendix 1: Importing and Exporting ASCII Files in ArcGIS............................17
Notes 18

Chapter 2 Measuring Distances and Time 19

2.1   Measures of Distance...19
2.2   Computing Network Distance and Time 21
    2.2.1   Label-Setting Algorithm for the Shortest-Route Problem 21
    2.2.2   Measuring Network Distance or Time in ArcGIS 23
2.3   Case Study 2: Measuring Distance between Counties and
Major Cities in Northeast China...24
    2.3.1   Part 1: Measuring Euclidean and Manhattan Distances 24
    2.3.2   Part 2: Measuring Travel Distances...26
    2.3.3   Part 3: Measuring Travel Time (Optional) 31
2.4   Summary 31
Appendix 2: The Valued-Graph Approach to the Shortest-Route Problem 31
Notes 33

Chapter 3 Spatial Smoothing and Spatial Interpolation.................................35

3.1   Spatial Smoothing 35
    3.1.1   Floating Catchment Area Method 36
    3.1.2   Kernel Estimation..37

3.2   Case Study 3A: Analyzing Tai Place-Names in Southern China by
      Spatial Smoothing 38
      3.2.1   Part 1: Spatial Smoothing by the Floating Catchment Area Method ...38
      3.2.2   Part 2: Spatial Smoothing by Kernel Estimation 41
3.3   Point-Based Spatial Interpolation 42
      3.3.1   Global Interpolation Methods 42
      3.3.2   Local Interpolation Methods 43
3.4   Case Study 3B: Surface Modeling and Mapping of Tai Place-Names in
      Southern China 45
      3.4.1   Part 1: Surface Mapping by Trend Surface Analysis 45
      3.4.2   Part 2: Mapping by Local Interpolation Methods 46
3.5   Area-Based Spatial Interpolation 47
3.6   Case Study 3C: Aggregating Data from Census Tracts to
      Neighborhoods and School Districts in Cleveland, Ohio 48
      3.6.1   Part 1: Simple Aggregation from Census Tracts to
              Neighborhoods in the City of Cleveland 49
      3.6.2   Part 2: Areal Weighting Aggregation from Census Tracts to
              School Districts in Cuyahoga County 49
3.7   Summary 51
Appendix 3: Empirical Bayes (EB) Estimation for Spatial Smoothing 52
Notes 53

PART II   Basic Quantitative Methods and Applications

Chapter 4   GIS-Based Trade Area Analysis and Applications in
            Business Geography and Regional Planning 55

4.1   Basic Methods for Trade Area Analysis 56
      4.1.1   Analog Method and Regression Model 56
      4.1.2   Proximal Area Method 56
4.2   Gravity Models for Delineating Trade Areas 57
      4.2.1   Reilly's Law 57
      4.2.2   Huff Model 59
      4.2.3   Link between Reilly's Law and Huff Model 60
      4.2.4   Extensions to the Huff Model 61
      4.2.5   Deriving the β Value in the Gravity Models 62
4.3   Case Study 4A: Defining Fan Bases of Chicago Cubs and White Sox 63
      4.3.1   Part 1: Defining Fan Base Areas by the Proximal Area Method 65
      4.3.2   Part 2: Defining Fan Base Areas and Mapping Probability
              Surface by the Huff Model 66
      4.3.3   Discussion 68
4.4   Case Study 4B: Defining Hinterlands of Major Cities in
      Northeast China 68

4.4.1   Part 1: Defining Proximal Areas by Railroad Distances..................69
4.4.2   Part 2: Defining Hinterlands by the Huff Model 69
4.4.3   Discussion 71
4.5. Concluding Remarks 71
Appendix 4: Economic Foundation of the Gravity Model..............................73
Notes 75

Chapter 5   GIS-Based Measures of Spatial Accessibility and Application in
              Examining Health Care Access 77

5.1   Issues on Accessibility 77
5.2   The Floating Catchment Area Methods 79
      5.2.1   Earlier Versions of Floating Catchment Area Method 79
      5.2.2   Two-Step Floating Catchment Area (2SFCA) Method 80
5.3   The Gravity-Based Method 82
      5.3.1   Gravity-Based Accessibility Index 82
      5.3.2   Comparison of the 2SFCA and Gravity-Based Methods.................83
5.4   Case Study 5: Measuring Spatial Accessibility to Primary Care
      Physicians in the Chicago Region 84
      5.4.1   Part 1: Implementing the 2SFCA Method 85
      5.4.2   Part 2: Implementing the Gravity-Based Model 89
5.5   Discussion and Remarks...91
Appendix 5: A Property for Accessibility Measures 95
Notes 96

Chapter 6   Function Fittings by Regressions and Application in
              Analyzing Urban and Regional Density Patterns...........................97

6.1   The Density Function Approach to Urban and Regional Structures 97
      6.1.1   Studies on Urban Density Functions 97
      6.1.2   Studies on Regional Density Functions.......................................99
6.2   Function Fittings for Monocentric Models 101
      6.2.1   Four Simple Bivariate Functions 101
      6.2.2   Other Monocentric Functions 102
      6.2.3   GIS and Regression Implementations 102
6.3   Nonlinear and Weighted Regressions in Function Fittings........................105
6.4   Function Fittings for Polycentric Models..107
      6.4.1   Polycentric Assumptions and Corresponding Functions................107
      6.4.2   GIS and Regression Implementations 110
6.5   Case Study 6: Analyzing Urban Density Patterns in the Chicago Region ...110
      6.5.1   Part 1: Function Fittings for Monocentric Models
              (Census Tracts)..111
      6.5.2   Part 2: Function Fittings for Polycentric Models
              (Census Tracts)..115
      6.5.3   Part 3: Function Fittings for Monocentric Models (Townships) 116
6.6   Discussion and Summary...117

Appendix 6A: Deriving Urban Density Functions 120
   Mills–Muth Economic Model...120
   Gravity-Based Model 121
Appendix 6B: OLS Regression for a Linear Bivariate Model 121
Appendix 6C: Sample SAS Program for Monocentric Function Fittings 123
Notes 124

Chapter 7   Principal Components, Factor, and Cluster Analyses, and
               Application in Social Area Analysis...................................127

7.1   Principal Components and Factor Analysis....................................127
      7.1.1   Principal Components Factor Model.................................128
      7.1.2   Factor Loadings, Factor Scores, and Eigenvalues...........129
      7.1.3   Rotation 130
7.2   Cluster Analysis 131
7.3   Social Area Analysis 134
7.4   Case Study 7: Social Area Analysis in Beijing.............................135
7.5   Discussion and Summary..143
Appendix 7A: Discriminant Function Analysis.....................................145
Appendix 7B: Sample SAS Program for Factor and Cluster Analyses 146
Notes 147

PART III   Advanced Quantitative Methods and Applications

Chapter 8   Geographic Approaches to Analysis of Rare Events in Small
               Population and Application in Examining Homicide Patterns 149

8.1   The Issue of Analyzing Rare Events in a Small Population.....................149
8.2   The ISD and the Spatial-Order Methods...150
8.3   The Scale-Space Clustering Method 152
8.4   Case Study 8: Examining the Relationship between Job Access and
      Homicide Patterns in Chicago at Multiple Geographic Levels Based
      on the Scale-Space Melting Method 155
8.5   Summary 163
Appendix 8: The Poisson-Based Regression Analysis 164
Notes 165

Chapter 9   Spatial Cluster Analysis, Spatial Regression, and Applications in
               Toponymical, Cancer, and Homicide Studies 167

9.1   Point-Based Spatial Cluster Analysis 168
      9.1.1   Point-Based Tests for Global Clustering 168
      9.1.2   Point-Based Tests for Local Clusters 168
9.2   Case Study 9A: Spatial Cluster Analysis of Tai Place-Names in
      Southern China...170

9.3   Area-Based Spatial Cluster Analysis 172
        9.3.1   Defining Spatial Weights 172
        9.3.2   Area-Based Tests for Global Clustering................................172
        9.3.3   Area-Based Tests for Local Clusters 173
9.4   Case Study 9B: Spatial Cluster Analysis of Cancer Patterns in Illinois 175
9.5   Spatial Regression 181
9.6   Case Study 9C: Spatial Regression Analysis of Homicide Patterns
        in Chicago 182
        9.6.1   Part 1: Spatial Regression Analysis at the Census Tract
                Level by GeoDa 183
        9.6.2   Part 2: Spatial Regression Analysis at the Community Area
                Level by GeoDa 185
        9.6.3   Discussion 185
9.7   Summary 187
Appendix 9: Spatial Filtering Methods for Regression Analysis 187
Notes 188

Chapter 10 Linear Programming and Applications in Examining Wasteful
               Commuting and Allocating Health Care Providers........................189

10.1 Linear Programming (LP) and the Simplex Algorithm 190
      10.1.1 The LP Standard Form 190
      10.1.2 The Simplex Algorithm 190
10.2 Case Study 10A: Measuring Wasteful Commuting in Columbus, Ohio 193
      10.2.1 The Issue of Wasteful Commuting and Model Formulation 193
      10.2.2 Data Preparation in ArcGIS 194
      10.2.3 Measuring Wasteful Commuting in SAS 197
10.3 Integer Programming and Location-Allocation Problems 199
      10.3.1 General Forms and Solutions 199
      10.3.2 Location-Allocation Problems 200
10.4 Case Study 10B: Allocating Health Care Providers in
      Cuyahoga County, Ohio..203
      10.4.1 Part 1: Polygon-Based Analysis 203
      10.4.2 Part 2: Network-Based Analysis..207
10.5 Discussion and Summary..212
Appendix 10A: Hamilton's Model on Wasteful Commuting..............................213
Appendix 10B: SAS Program for the LP Problem of Measuring
              Wasteful Commuting 214
Notes 217

Chapter 11 Solving a System of Linear Equations and Application in
               Simulating Urban Structure 219

11.1 Solving a System of Linear Equations...219
11.2 The Garin–Lowry Model 221
      11.2.1 Basic vs. Nonbasic Economic Activities......................................221
      11.2.2 The Model's Formulation 222
      11.2.3 An Illustrative Example 224

11.3 Case Study 11: Simulating Population and Service Employment
       Distributions in a Hypothetical City..225
       11.3.1 Task 1: Computing Network Distances (Times) in ArcGIS 226
       11.3.2 Task 2: Simulating Distributions of Population and
              Service Employment in the Basic Case 227
       11.3.3 Task 3: Examining the Impact of Basic Employment Pattern........229
       11.3.4 Task 4: Examining the Impact of Travel Friction Coefficient 229
       11.3.5 Task 5: Examining the Impact of the Transportation Network 230
11.4 Discussion and Summary...230
Appendix 11A: The Input–Output Model...231
Appendix 11B: Solving a System of Nonlinear Equations 232
Appendix 11C: FORTRAN Program for Solving the Garin–Lowry Model........234

References 243

Index..253

Related Titles...265

Part I

GIS and Basic Spatial Analysis Tasks

1 Getting Started with ArcGIS: Data Management and Basic Spatial Analysis Tools

A *Geographic Information System* (GIS) is a computer system that captures, stores, manipulates, queries, analyzes, and displays geographically referenced data. Among the diverse set of tasks a GIS can do, mapping remains the primary function. The first objective of this chapter is to demonstrate how GIS is used as a computerized mapping tool. The key skill involved in this task is the management of spatial and aspatial (attribute) data and the linkage between them. However, GIS is beyond mapping and has been increasingly used in various spatial analysis tasks as the GIS software becomes more capable and also friendlier to use for these tasks. The second objective of this chapter is to introduce some basic spatial analysis tools in GIS.

Given its wide use in education, business, and governmental agencies, ArcGIS is chosen as the major software platform to implement GIS tasks in this book. Unless pointed out otherwise, all studies in this book are based on ArcGIS 9.0. All chapters are structured similarly, beginning with conceptual discussions to lay out the foundation for methods, followed by case studies to get readers acquainted with techniques. Section 1.1 offers a quick tour of spatial and attribute data management in ArcGIS. Section 1.2 illustrates the typical process of GIS-based mapping in case study 1A: mapping the population density pattern in Cuyahoga County, Ohio. Section 1.3 surveys basic spatial analysis tools in ArcGIS, including queries, spatial joins, and map overlays. Section 1.4 illustrates some of the spatial analysis tools in case study 1B: extracting census tracts in Cleveland and deriving a polygon adjacency matrix. The polygon adjacency matrix defines spatial weights often needed in advanced spatial statistical studies such as spatial cluster and regression analyses (see Chapter 9). The chapter is concluded with a brief summary in Section 1.5.

It is assumed that readers have some basic GIS knowledge equivalent to one introductory GIS course. This chapter is not intended to review or cover all ArcGIS functions. Instead, it serves as a quick warm-up to prepare readers for more advanced spatial analysis in later chapters.

1.1 SPATIAL AND ATTRIBUTE DATA MANAGEMENT IN ArcGIS

Since ArcGIS is chosen as the primary software platform for this book, it is helpful to have a brief overview of its major modules and functions. *ArcGIS* was released in 2001 by Environmental Systems Research Institute, Inc. (ESRI) with a full-featured graphic user interface (GUI) to replace the older versions of *ArcInfo* based on typed commands. ArcGIS contains three major modules: ArcCatalog, ArcMap, and ArcToolbox. *ArcCatalog* views and manages spatial data files. *ArcMap* displays, analyzes, and edits the spatial data as well as attribute data. *ArcToolbox* contains various functions for managing and analyzing data, including managing map projections, converting between data formats, and implementing commands that are available from the older ArcInfo system. In ArcGIS 9.0, ArcToolbox is accessed in either ArcMap or ArcCatalog. Most, but not all, of the functions from the older ArcInfo system are available in the new ArcGIS system. For some tasks, we still need to access the old ArcInfo command interface. For example, Appendix 1 discusses how to import and export ASCII files in ArcInfo Workstation, and case study 2 (Section 2.3.2) uses ArcInfo Workstation to compute network distances.

1.1.1 MAP PROJECTIONS AND SPATIAL DATA MODELS

GIS differs from other information systems because of its unique capability of managing geographically referenced or spatial (location) data. Understanding the management of spatial data in GIS requires the knowledge of the *geographic coordinate system* with longitude and latitude values and its representations on various *plane coordinate systems* with *x, y* coordinates (map layers). Transforming the Earth's spherical surface to a plane surface or between different plane coordinate systems is a process referred to as *projection*. In ArcGIS, ArcMap automatically converts data of different coordinate systems to the first dataset added to the frame in map displays, commonly referred to as *on-the-fly reprojections*. However, this may be a time-consuming process if the dataset has a large size. It is a good practice to use the same projection for all data layers in one project. Two map projections are commonly used in the U.S.: *Universal Transverse Mercator* (UTM) and the *State Plane Coordinate System* (SPCS). Strictly speaking, the SPCS is not a projection; instead, it may use one of three different projections: Lambert Conformal Conic, Transverse Mercator, and Oblique Mercator. For the least distortion, north–south oriented states or regions use a Transverse Mercator projection, and east–west oriented states or regions use a Lambert Conformal Conic projection. Some states (e.g., Alaska, New York) may use more than one projection. For more details, one may refer to the Understanding Map Projections PDF file on the ArcGIS CD-ROM from the ESRI.

To check the existing projection for a spatial dataset in ArcGIS, one may use ArcCatalog by clicking the layer > choose Metadata > Spatial, or use ArcMap by right-clicking the layer > Properties > Source.

Projection-related tasks are conducted under ArcToolbox > Data Management Tools > Projections and Transformations. Under Projections and Transformations, the Define Projection tool creates a new projection definition file (PRJ) containing the projection parameters, or modifies an existing one (if incorrect). The Define Projection tool only labels a dataset with its correct coordinate system; it does not change the coordinates. If the spatial data are vector based, choose Feature > Project (also under Projections and Transformations) to actually transform the coordinate systems from one projection to another and generate a new layer. The tool provides the option to create a new coordinate system, use a predefined coordinate system, or import a coordinate system from an existing geodataset. If the spatial data are raster based, choose Raster > Project Raster.

Traditionally, a GIS uses either the vector or the raster data model to manage spatial data. A *vector GIS* uses geographically referenced points to construct spatial features of points, lines, and areas, and a *raster GIS* uses grid cells in rows and columns to represent spatial features. The raster data structure is simple and relatively easier to model. The vector data structure is used in most socioeconomic applications and is used in most applications in this book. Most commercial GIS software can convert from vector to raster data, or vice versa. In ArcGIS, the tools are available under ArcToolbox > Conversion Tools.

Earlier versions of ESRI's GIS software used the *coverage* data model. Later, *shapefiles* were developed for the ArcView package. Since the release of ArcGIS 8, the *geodatabase* model has become available and represents the new trend of object-oriented data model. The object-oriented data model stores the geometries of objects (spatial data) also as attribute data, whereas the traditional coverage or shapefile model stores spatial and attribute data separately. In any case, spatial and attribute data in socioeconomic analysis often come from different sources, and a typical task is to join them together in GIS for mapping and analysis. This involves attribute data management as discussed below.

1.1.2 Attribute Data Management and Attribute Join

A GIS includes both spatial and attribute data. Spatial data capture the geometry of map features, and attribute data describe the characteristics of map features. Attribute data are usually stored as a tabular file or table. ArcGIS reads several types of table formats. Shapefile attribute tables use the *dBase* format, ArcInfo Workstation uses the *INFO* format, and geodatabase tables use the *Microsoft Access* format. ArcGIS can also read several text formats, including comma-delimited text and tab-delimited text. Appendix 1 discusses how to import and export data in ASCII format in ArcGIS. Both tasks are important for advanced analysts who use GIS along with other software (e.g., SAS) or write programs for more complex computational tasks.

For basic tasks of data management, some can be done in both ArcCatalog and ArcMap, and some can be done only in ArcCatalog or ArcMap. Creating a new table or deleting/copying an existing table is done only in ArcCatalog (recall that Arc-Catalog is for viewing and managing GIS data files). A table is created in ArcCatalog by right-clicking the folder where the new table will be placed > New. A table can be deleted or copied in ArcCatalog by right-clicking the table > Delete (or Copy).

Adding a new variable to a table (either a "field" in a shapefile attribute table or dBase file or an "item" in an ArcInfo Workstation INFO file) can be done in both ArcCatalog and ArcMap. Deleting an existing item in an INFO file can be also done in both ArcCatalog and ArcMap, but deleting a field in a dBase file can only be done in ArcMap. For example, for adding a field to a shapefile attribute table, one may use ArcCatalog > right-click the shapefile > Properties > click Fields and type the new field name into an empty row in the Field Name column and define its Data Type. One may also add a field in ArcMap by opening the table > Options > Add Field.

TABLE 1.1
Types of Relationships in Combining Tables

Relationship	Match		Join or Relate in ArcGIS
One to one	One record in the destination table	One record in the source table	Join
Many to one	Multiple records in the destination table	One record in the source table	Join
One to many	One record in the destination table	Multiple records in the source table	Relate
Many to many	Multiple records in the destination table	Multiple records in the source table	Relate

For deleting a field in ArcMap, one needs to open the table > right-click the field > Delete Field. Updating values in a table is done in ArcMap: open the table > right-click the field and choose Calculate Values. In addition, in ArcMap, basic statistics for a field can be obtained by right-clicking the field and choosing Statistics.

In GIS, an *attribute join* is often used to link information in two tables based on a common field (key). The table can be an *attribute table* associated with a particular geodataset or a *stand-alone table*. In an attribute join, the field name of the key does not need to be identical in the two tables to be combined, but the data type must match. There are various relationships between tables in a join: one to one, many to one, one to many, and many to many. For either a one-to-one or many-to-one relationship, a *join* is used to combine two tables in ArcGIS. However, when the relationship is one to many or many to many, a join cannot be performed. ArcGIS uses a *relate* to link two tables while keeping the two tables separated. In a relate, one or more records are selected in one table, and the associated records are selected in another table. Table 1.1 summarizes the relationships and corresponding tools in ArcGIS.

A join or relate is accessed in ArcMap. Under Table of Contents, right-click the spatial dataset or the table that is to become the destination table, and choose Joins and Relates > Join (or Relate) > choose "Join attributes from a table" in the Join Data dialog window. A join is temporary in the sense that no new data are created and the join relationship is lost once the project is exited if the active project is not saved. The result of a join can be preserved by exporting the combined table to a new table.

Once the attribute information is joined to a spatial layer, mapping is convenient in ArcGIS. In ArcMap, right-click the layer and choose Properties > Symbology to invoke the dialog window. In the dialog, one can select a field to map, choose colors and symbols, and plan the layout. Map elements (scale, north arrow, legends) can be added to the draft map by clicking Insert from the main menu bar.

1.2 CASE STUDY 1A: MAPPING THE POPULATION DENSITY PATTERN IN CUYAHOGA COUNTY, OHIO

For readers without much exposure to GIS, nothing demonstrates the value of GIS and eases the fear of GIS complexity better than the experience of mapping data from the public domain with a few clicks. This section uses a case study to illustrate how spatial and aspatial information are managed and linked in a GIS, and how the information is used for mapping. The procedures are designed in a such a way that most functions introduced in Section 1.1 will be utilized.

A GIS project begins with data acquisition. In many cases, it uses existing data. For socioeconomic applications in the U.S., the *Topologically Integrated Geographical Encoding and Referencing* (TIGER) files from the Census Bureau are the major source for spatial data, and the decennial census data from the same agency are the major source for attribute data. Both can be accessed at the website www.census.gov. Advanced ArcGIS users may download the TIGER data directly and use the TIGER conversion tool to extract needed spatial data. The TIGER conversion tool can be accessed under ArcToolbox > Coverage Tools > Conversion > To Coverage > Advanced Tiger Conversion (or Basic Tiger Conversion). The conversion process may be time-consuming, and resulting files may require further editing work. Fortunately, many TIGER-based spatial layers are already processed and available in ArcGIS formats (shapefiles or coverage) in public domains. Some are contained in the data CDs that come with the ArcGIS software from the ESRI. If the spatial data are in the coverage interchange format (e00), the file can be converted to a coverage following ArcToolbox > Coverage Tools > Conversion > To Coverage > Import from Interchange File. In this case study, the spatial data are downloaded in shapefiles from the ESRI website.

Although readers can download the data as instructed, all data needed for the project are provided in the enclosed CD for convenience:

1. Shapefile `tgr39035trt00`
2. dBase file `tgr39000sf1trt.dbf`

Throughout this book, all computer file names and variable names (also some command lines that are specific to the projects) are in the `Courier New` font.

The following is a step-by-step guide for the process:

1. *Downloading spatial data*: Visit the ESRI's website for the Census 2000 TIGER/Line Data at http://www.esri.com/data/download/census2000_tigerline/. Select the state (Ohio) and the county (Cuyahoga) and download the census tract 2000 layer shapefile in WinRAR ZIP file. The unzipped shapefiles[1] share the name `tgr39035trt00`.[2]
2. *Transforming to UTM*: In ArcCatalog, we may check the projection for the shapefile `tgr39035trt00` and find out that it uses the geographic coordinate system. In ArcToolbox, choose Data Management Tools > Projections and Transformations > Feature > Project to invoke the dialog. In the dialog, select `tgr39035trt00.shp` as the Input Dataset, name the Output Dataset `cuyautm.shp`, and define the Output Coordinate System as UTM (zone 17; units, meters). Here we import the definition file from an existing dataset to define the output coordinate system: click the graphic icon under "Output Coordinate System" to activate the Spatial Reference Properties dialog > select Import > choose `clevbnd`. Figure 1.1 shows the dialog windows for the task. Click OK in both windows to execute.

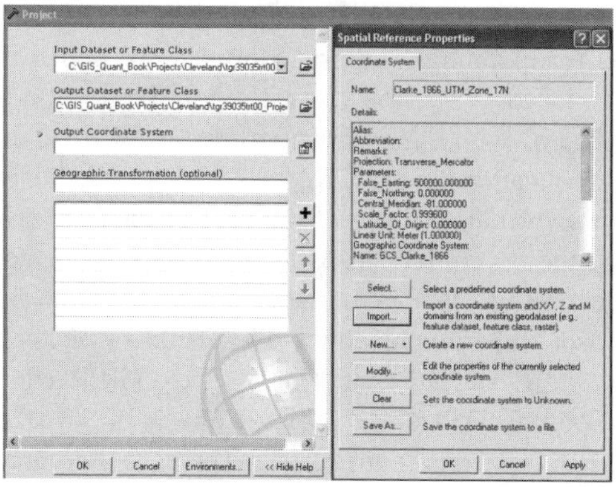

FIGURE 1.1 Dialog windows for projecting a spatial dataset.

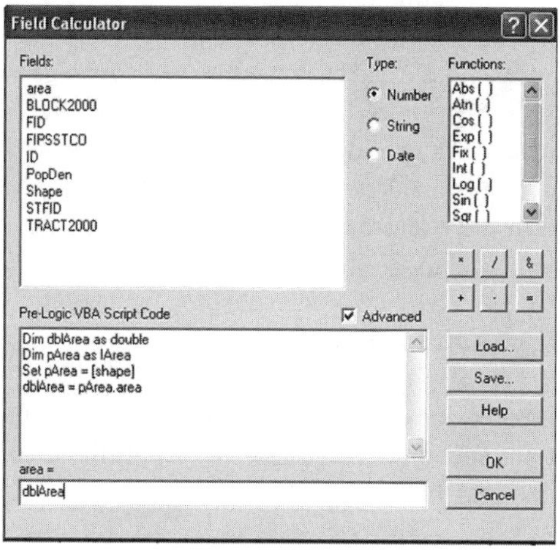

FIGURE 1.2 Dialog window for updating area in shapefile.

3. *Updating area for the shapefile*[3]: In ArcMap,[4] open the attribute table of cuyautm and add a new field area by clicking the Options button, selecting Add Field, and choosing the data type Double. Right-click the field area and choose Calculate Values. In the dialog, check Advanced and type the following Visual Basic (VBA) statements in the first text box:

```
Dim dblArea as double

Dim pArea as IArea

Set pArea = [shape]

dblArea = pArea.area
```

Type dblArea in the text box directly under the field area. Click OK to update areas. Figure 1.2 shows a sample dialog window for updating the areas.

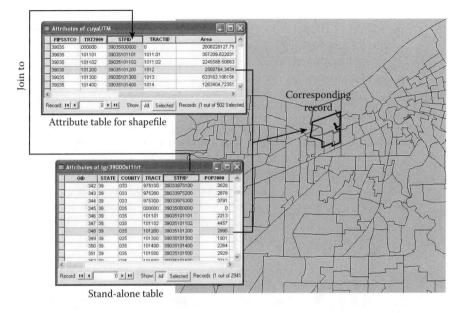

FIGURE 1.3 Attribute join in ArcGIS.

4. *Downloading attribute data*: The corresponding attribute dataset is down-loaded at the same site by selecting the layer "2000 census tract demo-graphics (SF1)." The attribute dataset `tgr39000sf1trt.dbf` is in dBase format for the whole state of Ohio, where `SF1` stands for Summary File 1 (i.e., the 100% count census data based on the short form).[5]

5. *Extracting attribute data for Cuyahoga County*[6]: In ArcMap, add the census data `tgr39000sf1trt.dbf` and open it. Click the Options tab at the lower right corner of the table > choose Select By Attributes > input the Structured Query Language (SQL) statement `county='035'` and click Apply (see Section 1.3 for more discussion on query in ArcGIS). All records in Cuyahoga County are selected. Click the Options tab again > Export. Name the extracted data file `cuya2k_popu.dbf`. If desirable, the table may be simplified by deleting all fields except for `STFID` and `POP2000` needed for the project.

6. *Joining spatial and attribute data*: Right-click the layer `cuyautm` > choose Joins and Relates > Join to join the table `cuya2k_popu.dbf` to the shapaefile `cuyautm` based on the common key `STFID`. STFID is the unique ID for each tract containing the codes for the state (two digits), county (three digits), and tract (six digits). Figure 1.3 shows how the spatial and attribute data are joined and related to the graphic elements on the map.

7. *Adding and calculating population density*: Right-click the layer `cuyautm` again and choose Open Attribute Table to examine the resulting combined table. In the combined table, a field name is now identified by its source table name and the original field name. For example,

cuyautm.area represents the field area from the attribute table of
shapefile cuyautm, and tgr39000sf1trt.STFID represents the
field STFID from the table tgr39000sf1trt.dbf (some long field
names may be truncated). In project instructions hereafter in this book,
we omit the source table name in most cases when referring to field names
(e.g., in calculation formulas, table joins, and others) unless we wish to
emphasize the source table.

Click the Options tab > Add Field to add a new field popuden. The
added field will be placed in the combined table as the last field of
cuyautm, but ahead of the first field of cuya2k_popu.dbf. Right-
click the field popuden, choose Calculate Values, and input the
formula 1000000*[POP2000]/[area]. In the formula, both fields
POP2000 and area are entered by clicking the field names from the
top box to save time and minimize chances of errors. In project instructions hereafter in this book, we simply write the formula such as
popuden=1000000*POP2000/area for similar tasks. Note the
map projection unit is meter, and the formula calculates population
densities as persons per square kilometer.

8. *Mapping population density pattern*: Right-click the layer cuyautm >
   choose Properties > Symbology > Quantities > Graduated Colors to map
   the field popuden. Experiment with different classification methods,
   number of classes, and color schemes. From the main menu bar, choose
   View > Layout View to preview the map. Also from the main menu bar,
   choose Insert > Legend (Scale Bar, North Arrow, and others) to add
   map elements.

   A population density map for the study area is shown in Figure 1.4. The
   large tract on the north is Lake Erie. The map uses customized breaks for
   density classes.

1.3 SPATIAL ANALYSIS TOOLS IN ArcGIS: QUERIES, SPATIAL JOINS, AND MAP OVERLAYS

Many spatial analysis tasks utilize the information on how spatial features are related
to each other in terms of the location. Spatial operations such as queries, spatial
joins, and map overlays[7] provide basic tools in conducting such tasks.

Queries include attribute (aspatial) queries and spatial queries. An *attribute query*
uses information in an attribute table to find attribute information (in the same table)
or spatial information (features in a spatial data layer). Attribute queries are accessed
in ArcMap: either (1) under Selection from the main menu bar > choose the option
Selection by Attributes or (2) in an opened table, choose the Options tab > Selection
by Attributes. Both allow users to select spatial features based on a query expression
in SQL using attributes (or to simply select attribute records from a stand-alone
table). Step 5 in case study 1A already used this function. Another option under
Selection in the main menu bar is Interactive Selection Method, which uses a pointer
to select features on the screen (in either a map or a table).

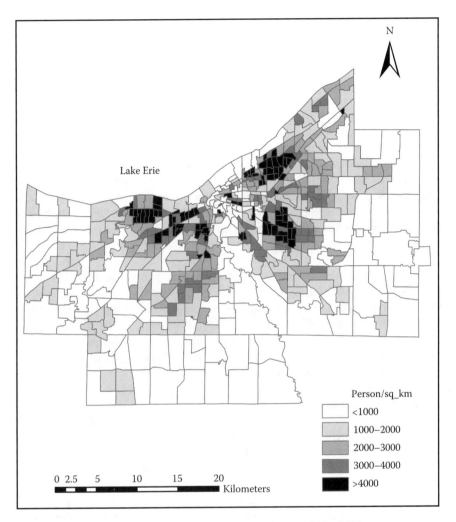

FIGURE 1.4 Population density pattern in Cuyahoga County, Ohio, 2000.

Compared to other information systems, one unique feature of GIS is its capacity of *spatial query*, which finds information based on location relationship between features from different layers. The option Selection by Location under Selection in the main menu bar searches for features in one layer based on its spatial relationship with another layer. The spatial relationships are defined by "intersect," "are within a distance of," "completely contain," "are completely within," etc.

Selected data by either an attribute query or a spatial query can be exported to a new dataset: (1) spatial features are saved in a layer by right-clicking the source layer and choosing Data > Export Data, and (2) attribute data are saved in a table by clicking the Options tab in the opened table > Export.

While an attribute join utilizes a common field between two tables, a *spatial join* uses the locations of spatial features, such as overlapping or proximity between two layers. We use the terms *source layer* and *destination layer* to differentiate the two layers by their roles: attributes from the source layer are processed and transferred to the destination layer. If one object in the source layer corresponds to one or multiple objects in the destination layer, it is a *simple join*. For example, by spatially joining a polygon layer of counties (source layer) with a point layer of school locations (destination layer), attributes of each county (e.g., county code, name, administrator) are assigned to schools that fall within the county boundary. If multiple objects in the source layer correspond to one object in the destination layer, two operations may be performed: summarized join and distance join. A *summarized join* summarizes the numeric attributes of features in the source layer (e.g., average, sum, minimum, maximum, standard derivation, or variance) and adds the information to the destination layer. A *distance join* identifies the closest object (out of many) in the source layer from the matching object in the destination layer and transfers all attributes of this nearest object (plus a distance field showing how close they are) to the destination layer. For example, one may spatially join a point layer of geocoded crime incidents (source layer) with a polygon layer of census tracts (destination layer), and generate aggregated crime counts in census tracts (summarized join); or spatially join a point layer of bus stations (source layer) with a point layer of census block centroids (destination layer) and identify the nearest bus stop to each block (distance join).

There are a variety of spatial joins between different spatial features (Price, 2004, pp. 287–288). Table 1.2 summarizes all types of spatial joins in ArcGIS. Spatial joins are accessed in ArcMap in a manner similar to that of attribute joins: right-click the source layer > choose Joins and Relates > Join. In the Join Data dialog window, choose "Join data from another layer based on spatial location," instead of "Join attributes from a table."

Map overlays may be broadly defined as any spatial analysis involving modifying features from different layers. The following reviews some of the most commonly used map overlay tools (available with any ArcGIS license): Clip, Intersect, Union, Buffer, and Multiple Ring Buffer. A *Clip* truncates the features from one layer using the outline of another. An *Intersect* overlays two layers and keeps only the areas that are common to both. A *Union* also overlays two layers but keeps all the areas from both layers. A *Buffer* creates areas by extending outward from point, line, or polygon features over a specified distance. A *Multiple Ring Buffer* generates buffer features based on a set of distances. In ArcGIS 9.0, the above map overlay tools are grouped under different tool sets through ArcToolbox > Analysis Tools: Clip is under the Extract tool set, Intersect and Union are under the Overlay tool set, and Buffer and Multiple Ring Buffer are grouped under the Proximity tool set. Other map overlay tools used in the projects of this book include Erase (Section 1.4.2, step 3), Near (Section 2.3.2, step 2), Point Distance (Section 2.3.1, step 2), Dissolve (Section 4.3.1, step 2), and Append (Section 4.3).

One may notice the similarity among spatial queries, spatial joins, and map overlays. Indeed, many spatial analysis tasks may be accomplished by any one of the three. Table 1.3 summarizes the differences between them. A spatial query only finds the information on screen and does not create new datasets (unless one chooses to export the selected records or features). A spatial join always saves the result in a new layer. There is an important difference between spatial joins and map overlays.

TABLE 1.2
Types of Spatial Joins in ArcGIS

Source Layer (S)	Destination Layer (D)	Simple Join	Distance Join	Summarized Join
Point	Point		For each point in D, find its closest point in S, and transfer attributes of that closest point to D	For each point in D, find all the points in S closer to this point than to any other point in D, and transfer the points' summarized attributes to D
Line	Point		For each point in D, find the closest line in S, and transfer attributes of that line to D	For each point in D, find all lines in S that intersects it, and transfer the lines' summarized attributes to D
Polygon	Point	For each point in D, find the polygon in S containing the point, and transfer the polygon's attributes to D	For each point in D, find its closest polygon in S, and transfer attributes of that polygon to D	
Point	Line		For each line in D, find its closest point in S, and transfer the point's attributes to D	For each line in D, find all the points in S that either intersect or lie closest to it, and transfer the points' summarized attributes to D
Line	Line	For each line in D, find the line in S that it is part of, and transfer the attributes of source line to D		For each line in D, find all the lines in S that intersect it, and transfer the summarized attributes of intersected lines to D
Polygon	Line	For each line in D, find the polygon in S that it falls completely inside, and transfer the polygon's attributes to D	For each line in D, find the polygon in S that it is closest to, and transfer the polygon's attributes to D	For each line in D, find all the polygons in S crossed by it, and transfer the polygons' summarized attributes to D
Point	Polygon		For each polygon in D, find its closest point in S, and transfer the point's attributes to D	For each polygon in D, find all the points in S that fall inside it, and transfer the points' summarized attributes to D
Line	Polygon		For each polygon in D, find its closest line in S, and transfer the line's attributes to D	For each polygon in D, find all the lines in S that intersect it, and transfer the lines' summarized attributes to D
Polygon	Polygon	For each polygon in D, find the polygon in S that it falls completely inside, and transfer the attributes of source polygon to D		For each polygon in D, find all the polygons in S that intersect it, and transfer the summarized attributes of intersected polygons to D

TABLE 1.3
Comparison of Spatial Query, Spatial Join, and Map Overlay

Basic Spatial Analysis Tools	Function	Whether a New Layer Is Created	Whether New Features Are Created	Computation Time
Spatial query	Finds information based on location relationship between features from different layers and displays on screen	No (unless the selected features are exported to a new dataset)	No	Least
Spatial join	Identifies location relationship between features from different layers and transfers the attributes to destination layer	Yes	No	Between
Map overlay	Overlays layers to create new features and saves the result in a new layer	Yes	Yes (splitting, merging, or deleting features to create new)	Most

A spatial join merely identifies the location relationship between spatial features of input layers and does not change existing spatial features or create new features. In the process of a map overlay, some of the input features are split, merged, or dropped for generating a new layer. In general, map overlay operations take more computation time than spatial joins, and spatial joins take more time than spatial queries.

1.4 CASE STUDY 1B: EXTRACTING CENSUS TRACTS IN THE CITY OF CLEVELAND AND ANALYZING POLYGON ADJACENCY

The following datasets are needed for the project and are provided in the CD:

1. Shapefile `cuyautm` contains the census tracts in Cuyahoga County, Ohio.
2. Coverage `clevbnd` defines the boundary for the city of Cleveland, OH.

Each coverage in the CD is in ArcInfo interchange file (.e00) format, which needs to be converted back for use in ArcToolbox by selecting Coverage Tools > To Coverage > Import From Interchange File. Dataset 1 is a product from case study 1A, but provided in the CD so that one may start this project independent of case study 1A. The coverage `clevbnd` for the city boundary is found in a public domain.

1.4.1 PART 1: EXTRACTING CENSUS TRACTS IN CLEVELAND

In many cases, GIS analysts need to extract a study region that is contained in a larger area. This part of the project extracts the census tracts in Cleveland from Cuyahoga

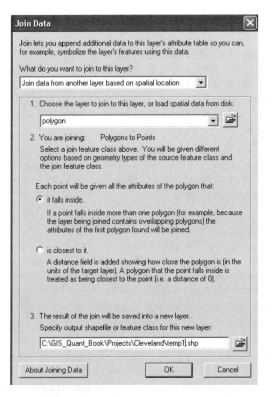

FIGURE 1.5 Dialog window for spatial join.

County. Overlaying the two layers shows that `clevbnd` is slightly off the census tract boundaries in `cuyautm` near the borders, and `cuyautm` contains more geographic details. Although the census tracts in `cuyautm` do not exactly match the boundary of `clevbnd`, their centroids are within the boundary of `clevbnd`. The plan is to identify the centroids of `cuyautm` that fall inside `clevbnd` and use this information to extract a polygon layer of census tracts for the city. If one simply uses `clevbnd` to clip `cuyautm`, not all geographic details of tracts in `cuyautm` would be preserved.

1. *Generating county census tract centroids*: Activate ArcToolbox in ArcMap > choose Data Management Tools > Features > Feature To Point. In the dialog, choose `cuyautm` as Input Features, name `cuya_pt` for Output Feature Class, and check the option Inside. A shapefile `cuya_pt` for all tract centroids is generated.
2. *Identifying tract centroids inside the city boundary*: Right-click the destination layer `cuya_pt` and choose Joins and Relates > Join. In the dialog, choose the option "Join data from another layer based on spatial location," the source layer "`clevbnd` polygon," and the option "it falls inside," and name the output shapefile `tmp1`. Figure 1.5 shows the dialog window for a spatial join. Open the attribute table for `tmp1` and note that `clevbnd_id` = 1 for all tracts inside the city boundary, 0 for those outside.

3. *Attaching information of identified tract centroids to the polygon layer*:
   Add the layer `cuyautm`, right-click it and choose Joins and Relates >
   Join > choose the option "Join attributes from a table" and `tmp1` as the
   source table, and use `STFID` as the common key (in both the destination
   layer `cuyautm` and the source table `tmp1`).
4. *Extracting census tracts in the city*: Open the attribute table of `cuyautm`
   > click the Options tab > choose Select by Attributes > input the selection
   criterion `tmp1.clevbnd_id = 1`. All polygons within the city are
   selected and highlighted. Right-click the layer `cuyautm` and choose
   Data > Export Data > make sure that "Selected features" is selected in the
   top box, and name the output `clevtrt`. The shapefile `clevtrt` contains
   all census tracts in Cleveland.

The above instructions use a spatial join. As explained in Section 1.3, one may
also use a spatial query tool (Selection by Location) or a map overlay tool
(ArcToolbox > Analysis Tools > Overlay > Identity) to accomplish the same task.
For example, a spatial query tool can be used to obtain the same result in one step:
choose Selection from the main menu bar > Selection by Location > use the dialog
to "select features from `cuyautm` that have their center in `clevbnd` polygon" >
export the selected features to a shapefile `clevtrt`.

1.4.2 PART 2: IDENTIFYING CONTIGUOUS POLYGONS

Deriving the *polygon adjacency matrix* is a very important task in spatial analysis.
For example, in Chapter 9, both the area-based spatial cluster analysis and spatial
regression utilize the matrix to define *spatial weights* in order to account for spatial
autocorrelation. Adjacency between polygons may be defined in two ways: (1) *rook
contiguity* uses only common boundaries to define adjacency, and (2) *queen
contiguity* includes all common points (boundaries and vertices) (Cliff and Ord,
1973). For the rook contiguity, one may simply use the ArcInfo Workstation command PALINFO to generate the polygon adjacency matrix. This case study uses the
queen contiguity to define polygon adjacency in order to illustrate some of the basic
spatial analysis tools discussed in Section 1.3.

Here we focus on identifying contiguous polygons for one exemplary tract.
Similar to step 4 in Part 1, select the tract with `TRACTID = '1038'` from the layer
`clevtrt` and export the selected feature to a shapefile `zonei`. This part of the
project is to identify neighboring tracts for `zonei` based on the queen contiguity.
Figure 1.6 shows the areas around the sample tract with their TRACTID values.
Based on the queen contiguity, there are six neighboring tracts (1026, 1028, 1029,
1035, 1036, and 1039) for tract 1038. If based on the rook contiguity, tract 1028
would not be included as a neighboring tract.

The following implements the task of identifying contiguous tracts for `zonei`:

1. *Buffering tract*: In ArcToolbox, choose Analysis Tools > Proximity >
   Buffer. Buffer a small distance (say, 30 m)[8] around the Input Features
   `zonei`, and name the Output Feature Class `zonei_buff`.

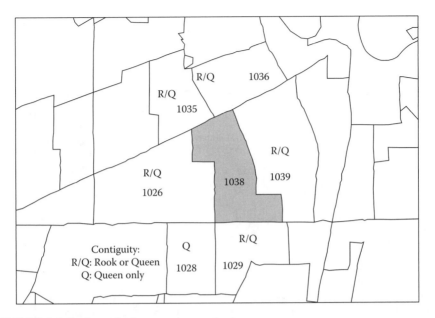

FIGURE 1.6 Rook contiguity vs. queen contiguity.

2. *Clipping the buffer from study area*: In ArcToolbox, choose Analysis Tools > Extract > Clip. Choose `clevtrt` as the Input Features and `zonei_buff` as the Clip Features, and name the Output Feature Class `zonei_clip`.
3. *Extracting neighboring polygons*: In ArcToolbox, choose Analysis Tools > Overlay > Erase. Choose `zonei_clip` as the Input Features and `zonei` as the Erase Features, and name the Output Feature Class `zonejs`. The shapefile `zonejs` contains all neighboring tracts of `zonei` based on the queen contiguity.

Figure 1.7 illustrates the process. The layer `zonei` is a single tract 1038. By buffering, the output layer `zonei_buff` contains a single polygon (the original tract is shown inside only for reference). By clipping the buffered zone from the study area, `zonei_clip` has seven polygons, including the original tract. The original tract `zonei` is removed by erasing it from `zonei_clip`. The output is a six-tract layer `zonejs` (with TRACTIDs shown in Figure 1.7).

Deriving a polygon adjacency matrix for the study area requires iterations of the same process on all tracts. An Arc Micro language (AML) program Queen_Cont.`aml` is provided in the CD for this task, which is based on Shen (1994).

At the end of the project, one may use ArcCatalog to delete unneeded data to save disk space, but keep the layers `cuyautm`, `cuya_pt`, and `clevtrt`, which will be used in future projects.

1.5 SUMMARY

The following summarizes major GIS and spatial analysis skills learned in this chapter:

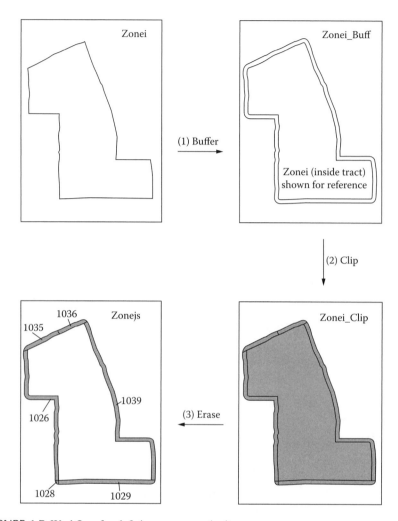

FIGURE 1.7 Workflow for defining queen contiguity.

1. Spatial data formats in ArcGIS and conversions between them
2. Map projections and transformations between them
3. Attribute data management (creating, editing, and deleting datasets and fields)
4. Attribute join (including joining attribute data with spatial data)
5. Mapping attributes
6. Spatial joins
7. Attribute and spatial queries
8. Map overlay operations (clip, buffer, intersect, union, and erase)

Other noticeable tasks include finding spatial and attribute data in the public domain, updating areas in a shapefile, and extracting centroids from a polygon layer to create a point layer. Projects in other chapters will also utilize these skills.

This chapter also discusses key concepts such as different relationships between tables (one to one, many to one, one to many, and many to many), various spatial joins, and differences among spatial queries, spatial joins, and map overlays.

For additional practice of GIS-based mapping, readers can download the census data and TIGER files for a familiar area and map some demographic (population, race, age, sex, etc.) and socioeconomic (income, poverty, educational attainment, family structure, housing characteristics, etc.) variables in the area.

APPENDIX 1: IMPORTING AND EXPORTING ASCII FILES IN ArcGIS

For a small dataset in ASCII (text) format, it is fairly easy to convert it to ArcGIS-readable formats. One may use Microsoft Excel to open the file, add a top row containing field headings, and save the file as a CSV file (comma-delimited text file). CSV files are readable in ArcGIS. One crucial shortcoming in this process is the lack of control on defining field types and formats. For example, all values for census tract codes or STFID in an ASCII file appear to be numeric. A CSV file generated from the ASCII file automatically defines the field as numeric and makes it infeasible to join the data to a GIS layer extracted from TIGER files, which usually defines the field as character.

Conversely, in ArcMap, one may open a table (either a feature attribute table or a stand-alone table) and choose Options > Export to export the table in dBase format. The dbase file can then be read in Microsoft Excel and saved as an ASCII file. Alternatively, one may also use ArcToolbox > Spatial Statistics Tools > Utilities > Export Feature Attribute to Ascii, which exports feature class coordinates and selected attribute values to a space-, comma-, or semicolon-delimited ASCII text file. However, only one variable (along with the coordinates) is exported at a time.

Microsoft Access is also a commonly used package for attribute data management, including tasks of importing and exporting ASCII files. The following discusses how to import and export a large ASCII file through ArcInfo Workstation.

1. *Converting ASCII to INFO file*:
    a. In ArcCatalog, click on the directory or workspace that is to contain the new file, choose File > New > INFO table to create a new INFO file (say, `ninfo`), and define all fields (name, data type, and format).
    b. Invoke the ArcInfo command interface in Windows by choosing Start > All Programs > ArcGIS > ArcInfo Workstation > Arc. Type `w ...` to navigate to the right workspace (e.g., `w c:\Quant_GIS\proj1`), and then type `tables` to activate the TABLES module.
    c. In TABLES, type `select ninfo` to choose the newly defined INFO file.
    d. Say `tfile` is the text file to be converted. Type `add from tfile` to append all data. The converted table can be viewed by typing the command `list`.
    e. Type `quit` to exit TABLES and type `quit` again to exit ArcInfo Workstation.

2. *Exporting INFO file or ArcGIS coverage feature tables to ASCII*: Conversely, an INFO file or an ArcInfo feature table (.PAT or .AAT) can be exported to an ASCII file as follows:
   a. In ArcInfo Workstation, navigate to the workspace and type `tables` to activate TABLES module.
   b. In TABLES, type `select ninfo` (e.g., `ninfo` is the file name) to choose the file.
   c. Type `unload tfile1` to export all items of the data to an ASCII file (say, named as `tfile1`). One may limit the fields to be exported by a command, `unload yfile item1 item2` (say, the field names to be unloaded are `item1 and item2`), or further specify the format by adding the option COLUMNAR, such as `unload yfile item1 item2 columnar`. The option COLUMNAR may be very useful as the output ASCII file is space delimited and each field takes exactly the same space as defined in the INFO file. The UNLOAD command without the option exports to a comma-delimited ASCII file and encloses values from character fields with single quotes ('').

NOTES

1. A layer in the shapefile format has at least three files (.dbf, .shp, .shx) associated with it. Some contain additional files (.prj, .sbx, .avl, .xml). For convenience, the remainder of this book uses the singular term *shapefile* to refer to multiple files associated with one shapefile layer.
2. In the file name, `tgr` indicates its source from TIGER files, 39 is the state's FIPS code, 035 is the county code, and `trt00` stands for census tracts in 2000.
3. Alternatively, one may use ArcToolbox > Spatial Statistics Tools > Utilities > Calculate Areas to implement the task (a new output feature will be created).
4. Unless specified otherwise, ArcMap is the work environment in all project instructions.
5. This dBase file does not contain all census variables. For complete census data, visit the 2000 Census website at http://www.census.gov/main/www/cen2000.html. Processing the files (e.g., SF1, SF3) requires understanding the 2000 Census data structure and using some data analysis software (e.g., SAS, Access).
6. One may omit this step and join the table `tgr39000sf1trt.dbf` directly to the layer `cuyautm`. The resulting layer automatically excludes records of other counties.
7. Some may consider these traditional GIS terms outdated. We use the terms to emphasize distinctive tasks accomplished by these spatial operations.
8. The buffer distance must be larger than the fuzzy tolerance (i.e., about 1 m for zonei), but small enough not to go beyond the immediate neighboring polygons.

2 Measuring Distances and Time

This chapter discusses one of basic tasks encountered most often in spatial analysis: measuring distances and time. After all, spatial analysis is about how physical and human activities vary across space — in other words, how these activities change with distances from reference locations or objects of interest. In many applications, once the distance or time measure is obtained, studies may be completed outside a GIS environment. The advancement and wide availability of GIS have made the task much easier than it used to be.

The task of distance or time estimation can be found throughout this book. For example, spatial smoothing and spatial interpolation in Chapter 3 utilize distance measures to determine which objects enter the computation and how much the objects influence the computation. In trade area analysis in Chapter 4, distances (or time) between stores and consumers dictate which stores are the closest and how often residents visit a store. In Chapter 5 on accessibility measures, distance or time measures are the building block of either the floating catchment area method or the gravity-based method. Chapter 6 examines how population density or land use intensity declines with distance from a city or regional center. The task can also be found in other chapters.

This chapter is structured as follows. Section 2.1 provides an overview of various distance measures. Section 2.2 discusses how to compute the shortest-route distance (time) through a network and how to implement it in ArcGIS. A case study of measuring the Euclidean and network distances in northeast China is presented in Section 2.3. Results from this case study will be used in case study 4B (Section 4.4). The chapter is concluded with a brief summary in Section 2.4.

2.1 MEASURES OF DISTANCE

Distance measures include Euclidean (straight-line, or air) distance, Manhattan distance, or network distance. *Euclidean distance* is simply the distance between two points through a straight line. Unless otherwise specified, distance is measured in Euclidean distance.

Prior to the wide usage of GIS, researchers needed to use mathematical formulas to compute the distance, and the accuracy is limited depending on the information available and tolerance of computational complexity. If a study area is small in terms of its geographic territory (e.g., a city or a county), Euclidean distance between two nodes (x_1, y_1) and (x_2, y_2) in Cartesian coordinates is approximated as

$$d_{12} = [(x_1 - x_2)^2 + (y_1 - y_2)^2]^{1/2} \tag{2.1}$$

If the study area covers a large territory (e.g., a state or a nation), one needs to compute the geodetic distance. The *geodetic distance* between two points is the distance through a great circle assuming the Earth as a globe. Given the geographic coordinates of two points as (a, b) and (c, d) in decimal degrees, the geodetic distance between them is

$$d_{12} = r * a \cos[\sin b * \sin d + \cos b * \cos d * \cos(c - a)] \tag{2.2}$$

where r is the radius of the earth (approximately 6367.4 km).

As the name suggests, Manhattan distance describes a rather restrictive movement in rectangular blocks, like in the borough of Manhattan. *Manhattan distance* is the length of the change in the x direction plus the change in the y direction. For instance, the Manhattan distance between two nodes (x_1, y_1) and (x_2, y_2) in Cartesian coordinates is simply computed as

$$d_{12} = | x_1 - x_2 | + | y_1 - y_2 | \tag{2.3}$$

Like Equation 2.1, Manhattan distance, defined by Equation 2.3, is only meaningful within a small study area (e.g., a city).

Network distance is the shortest-path (or least-cost) distance through a road network and will be discussed in detail in Section 2.2. Manhattan distance can be used as an approximation for network distance if the street network is in a grid pattern.

In ArcGIS, simply click on the graphic tool ⟨measure icon⟩ (measure) in ArcMap to obtain the Euclidean distance between two points (or a cumulative distance along several points). Distance is created as a by-product in many spatial analysis operations in ArcGIS. For example, a *distance join* (a spatial join method) in ArcGIS, as explained in Section 1.3, records the nearest distances between objects of two spatial datasets. In a distance join, distance between lines or polygons is between their closest points. Under ArcToolbox > Analysis Tools > Proximity, the Near tool computes the distance from each point in one layer to its closest polyline or point in another layer. Some applications need to use distances between any two points either within one layer or between different layers, and thus a distance matrix. The Point Distance tool in ArcToolbox is designed for this purpose and is accessed in ArcToolbox > Analysis Tools > Proximity > Point Distance. In the output file, if the value for DISTANCE is 0, it could be that the actual distance is either indeed 0 (e.g., from a point to itself) or beyond the Search radius.

The current ArcGIS version does not have a built-in tool for computing the less commonly used Manhattan distance. Computing Manhattan distances requires the Cartesian coordinates of points that can be generated in ArcToolbox. For a shapefile, use Data Management Tools > Features > Add XY Coordinates. For a coverage, use Coverage Tools > Data Management > Tables > Add XY Coordinates. Computing network distance in ArcGIS is more complex and will be discussed in the next two sections.

2.2 COMPUTING NETWORK DISTANCE AND TIME

A *network* consists of a set of *nodes* (or vertices) and a set of *arcs* (or edges or links) that connect the nodes. If the arcs are directed (e.g., one-way streets), the network is a *directed network*. A network without regard to direction may be considered a special case of directed network with each arc having two permissible directions. Finding the shortest chains from a specified origin to a specified destination is the *shortest-route problem*, which records the shortest distance or the least time (cost) if the impedance value (e.g., travel speed) is provided on each arc. Different methods for solving the problem have been proposed in the literature, including the label-setting algorithm discussed in this section and the valued-graph (or *L* matrix) method in Appendix 2.

2.2.1 LABEL-SETTING ALGORITHM FOR THE SHORTEST-ROUTE PROBLEM

The popular *label-setting algorithm* was first described by Dijkstra (1959). The method assigns labels to nodes, and each label is actually the shortest distance from a specified origin. To simplify the notation, the origin is assumed to be node 1. The method takes four steps:

1. Assign the *permanent* label $y_1 = 0$ to the origin (node 1) and a *temporary* label $y_j = M$ (a very large number) to every other node. Set $i = 1$.
2. From node i, recompute the temporary labels $y_j = \min (y_j, y_i + d_{ij})$, where node j is temporarily labeled and $d_{ij} < M$ (d_{ij} is the distance from i to j).
3. Find the minimum of the temporary labels, say, y_i. Node i is now permanently labeled with value y_i.
4. Stop if all nodes are permanently labeled; go to step 2 otherwise.

The following example is used to illustrate the method. Figure 2.1a shows the network layout with nodes and links. The number next to a link is the impedance value for the link.

Following step 1, permanently label node 1 and set $y_1 = 0$; temporarily label $y_2 = y_3 = y_4 = y_5 = M$. Set $i = 1$. A permanent label is marked with an asterisk (*). See Figure 2.1b.

In step 2, from node 1 we can reach nodes 2 and 3, which are temporarily labeled. $y_2 = \min (y_2, y_1 + d_{12}) = \min (M, 0 + 25) = 25$, and similarly, $y_3 = \min (y_3, y_1 + d_{13}) = \min (M, 0 + 55) = 55$.

In step 3, the smallest temporary label is $\min (25, 55, M, M) = 25 = y_2$. Permanently label node 2 and set $i = 2$. See Figure 2.1c.

Back to step 2, as nodes 3, 4, and 5 are still temporarily labeled. From node 2, we can reach temporarily labeled nodes 3, 4, and 5. $y_3 = \min (y_3, y_2 + d_{23}) = \min (55, 25 + 40) = 55$, $y_4 = \min (y_4, y_2 + d_{24}) = \min (M, 25 + 45) = 70$, $y_5 = \min (y_5, y_2 + d_{25}) = \min (M, 25 + 50) = 75$.

Following step 3 again, the smallest temporary label is $\min (55, 70, 75) = 55 = y_3$. Permanently label node 3 and set $i = 3$. See Figure 2.1d.

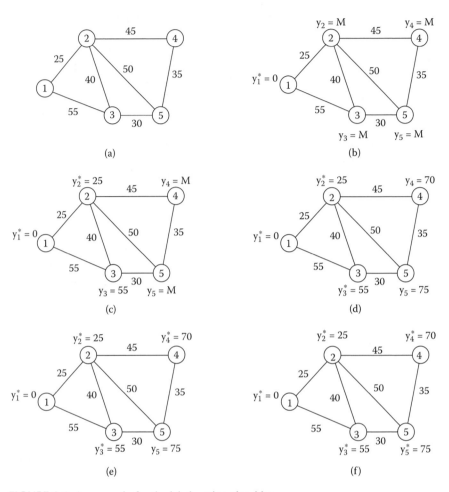

FIGURE 2.1 An example for the label-setting algorithm.

Back to step 2, as nodes 4 and 5 are still temporarily labeled. From node 3 we can reach only node 5 (still temporarily labeled). $y_5 = \min (y_5, y_3 + d_{35}) = \min (75, 55 + 30) = 75$.

Following step 3, the smallest temporary label is $\min (70, 75) = 70 = y_4$. Permanently label node 4 and set $i = 4$. See Figure 2.1e.

Back to step 2, as node 5 is still temporarily labeled. From node 4 we can reach node 5. $y_5 = \min (y_5, y_4 + d_{45}) = \min (75, 70 + 35) = 75$.

Node 5 is the only temporarily labeled node, so we permanently label node 5. By now all nodes are permanently labeled, and the problem is solved. See Figure 2.1f.

The permanent labels y_i give the shortest distance from node 1 to node i. Once a node is permanently labeled, we examine arcs "scanning" from it only once. The shortest paths are stored by noting the scanning node each time a label is changed (Wu and Coppins, 1981, p. 319). The solution to the above example can be summarized in Table 2.1.

TABLE 2.1
Solution to the Shortest-Route Problem

Origin–destination nodes	Arcs on the Route	Shortest distance
1, 2	(1, 2)	25
1, 3	(1, 3)	55
1, 4	(1, 2), (2, 4)	70
1, 5	(1, 2), (2, 5)	75

2.2.2 MEASURING NETWORK DISTANCE OR TIME IN ARCGIS

Networks handled in ArcGIS include transportation networks and utility networks. For our purpose, the discussion is limited to transportation networks. Most GIS textbooks (e.g., Chang, 2004, chap. 16; Price, 2004, chap. 14) discuss how the distance between two points (or distances between a location and many others) is obtained in ArcGIS. In many spatial analysis applications, a distance matrix between a set of origins and a set of destinations is needed. For this task, one needs to use the ArcInfo Workstation, in particular, the NODEDISTANCE command in the ArcPlot module. The NODEDISTANCE command computes the shortest distances through a road network by default and also outputs the Euclidean or Manhattan distances as options. By properly defining the item IMPEDANCE as time or cost, it also computes the shortest travel time or the least cost, respectively. The following explains how a matrix of network distances is computed in ArcGIS.

The first step is to set up the network. A transportation network has many network elements, such as link impedances, turn impedances, one-way streets, and overpasses and underpasses, that need to be defined (Chang, 2004, p. 351). Putting together a road network requires extensive data collection and processing, which can be very expensive or infeasible for many applications. For example, a road layer extracted from the TIGER/Line files does not contain nodes on the roads, turning parameters, or speed information. When such information is not available, one may assume that nodes built from a road layer by some automation tools (e.g., topology builders in ArcGIS) are acceptable and closely resemble the real-world network. For link impedances, one may assign speed limits based on road levels and account for congestion effects if possible. In Luo and Wang (2003), speeds are assigned to different roads according to the census feature class codes (CFCCs) used by the U.S. Census Bureau in its TIGER/Line files and whether in urban, suburban, or rural areas. Wang (2003) uses regression models to predict travel speeds by land use intensity (business and residential densities) and other factors.

In the second step, the NETCOVER command is used to set up the route system for network computation.

The third step is to define the origin nodes, destination nodes, and impedance item. Commands such as CENTERS, STOPS, and NODES are used to define origin and destination points; IMPEDANCE specifies which item in the network attribute table defines the impedance.

Finally, the NODEDISTANCE command is executed to calculate the network distances from origin nodes to destination nodes.

Note that the NODEDISTANCE command only computes the distances between nodes that are on the network. However, points of origins or destinations may not fall on the network. The distances between origins (destinations) and network nodes may be minor, but need to be included in the trips. This makes an important step in measuring network distances, as shown in case study 2 in the following section.

2.3 CASE STUDY 2: MEASURING DISTANCE BETWEEN COUNTIES AND MAJOR CITIES IN NORTHEAST CHINA

This case study measures distances between counties and major cities in northeast China. Results from this study will be used by case study 4B on defining urban hinterlands (see Chapter 4, Section 4.4).

The study area has been a relatively coherent region (i.e., the Northeast China Plain) for a long time. It includes three provinces: Heilongjiang, Jilin, and Liaoning. Based on their population and economic sizes, four major cities are identified: three provincial capitals (Harbin, Changchun, and Shenyang) and Dalin. As the railway remains the major mode for both passenger and freight transportation in China (even more so in the region), railroads are used for measuring network distances. See Figure 2.2 for the study area.

The following datasets are provided in the CD for the project:

1. Polygon coverage cntyne containing all 203 counties (or administrative units equivalent to county) in northeast China
2. Point coverage city4 containing four major cities in the region
3. Line coverage railne for railway network in the study area[1]

The railway network covers areas beyond the three provinces to maintain network connectivity.

2.3.1 PART 1: MEASURING EUCLIDEAN AND MANHATTAN DISTANCES

As explained earlier, both Euclidean and Manhattan distances may be obtained by choosing the options in the NODEDISTANCE command. In this part of the project, we compute these two measures without involving network analysis. As Manhattan distance is not an appropriate measure at a regional scale (see Section 2.1), the computation of Manhattan distances in steps 3 to 5 is only for demonstration and indicated as optional.

1. *Generating county centroids*: In ArcToolbox, choose Data Management Tools > Features > Feature To Point > choose cntyne as Input Features, name CntyNEpt for Output Feature Class (county centroids), and check the option Inside.
2. *Computing Euclidean distances*: In ArcToolbox, choose Analysis Tools > Proximity > Point Distance > choose CntyNEpt as Input Features and

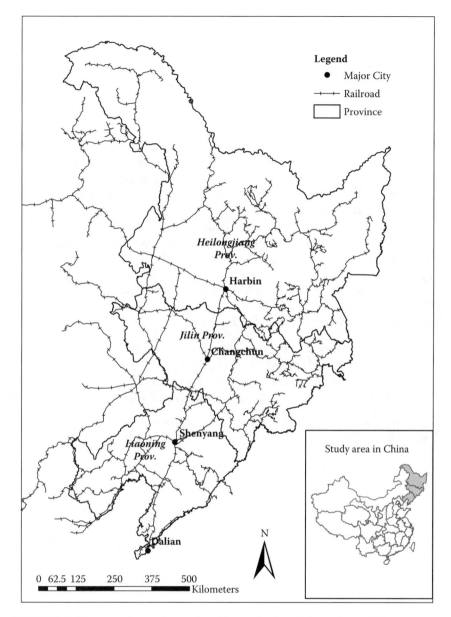

FIGURE 2.2 Three provinces, four major cities, and railroads in northeast China.

city4 (point) as Near Features, and name the output table Dist.dbf. There is no need to define a search radius, as all distances are needed. Note that there are 203 (counties) × 4 (cities) = 812 records in the distance table. Add a field airdist to the distance table and calculate it as airdist=distance/1000 to indicate that it is air (Euclidean) distance in kilometers (the projection unit is meter).

3. Optional: *Adding XY coordinates for county centroids and major cities*: In ArcToolbox, choose Data Management Tools > Features > Add XY Coordinates > choose `CntyNEpt` as Input Features. In the attribute table of `CntyNEpt`, results are saved in the fields `point-x` and `point-y`. Also in ArcToolbox, choose Coverage Tools > Data Management > Tables > Add XY Coordinates > choose `city4` as Input Coverage. In the attribute table of `city4`, results are saved in the fields `x-coord` and `y-coord`.

4. Optional: *Attaching coordinates to counties and cities in the distance table*: In ArcMap, right-click the table `Dist.dbf` > choose Joins and Relates > Join > use `FID` in `CntyNEpt` (source table) and `INPUT_FID` in `Dist.dbf` (destination table) as the common keys to join the two tables. Similarly, use `FID` in `City4` and `NEAR_FID` in the updated table `Dist.dbf` as the common keys to join them.

5. Optional: *Computing Manhattan distances*: Open the updated table `Dist.dbf`, add a field `Manhdist`, and calculate it as `Manhdist = abs(x-coord - point-x)/1000+abs(y-coord - point-y) /1000`. The computed Manhattan distances are in kilometers and are always larger than the corresponding Euclidean distances.

2.3.2 PART 2: MEASURING TRAVEL DISTANCES

The travel distance between an origin county and a destination city is composed of three segments. Figure 2.3 shows an example: (1) the first segment (S1) is the distance from county 76 to its closest node (171) on the road network, (2) the second segment (S2) is the network distance between nodes 171 and 162 through the

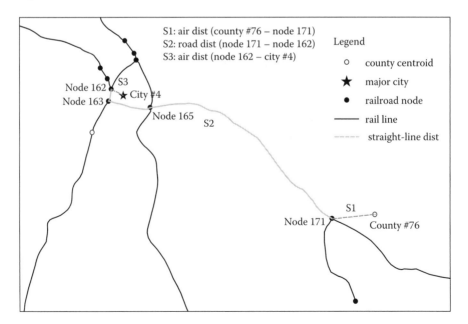

FIGURE 2.3 Three segments in measuring travel distance.

railroads (passing nodes 165 and 163), and (3) the third segment (S3) is the distance from city 4 to its closest node (162) on the road network. Segments S1 and S3 are approximated by straight-line (air) distances, and segment S2 is the network distance between nodes. In other words, from county 76 to city 4 it is assumed that one travels from county 76 to the nearest node (171), then travels through the railroads to 162 (passing nodes 165 and 163), and finally stops at city 4. The task in this part of the project is to find these nodes that are closest to counties and cities, compute these three distance segments, and finally sum them up.

1. *Preparing the network coverage*: In ArcToolbox, use Coverage Tools > Data Management > Topology > Build to build the line topology on the coverage `railne`. Repeat the process to build the node topology on it.[2]

2. *Computing air distances between counties/cities and their nearest nodes*: In ArcToolbox, choose Analysis Tools > Proximity > Near > choose `CntyNEpt` as Input Features and `railne` (node) as Near Features. In the updated attribute table for `CntyNEpt`, the field `NEAR_FID` identifies the closest node on the railway network to a county, and another field `NEAR_DIST` identifies the distance between them. To identify the nearest nodes from major cities, repeat the step on the coverage `city4`: choose `city4` (point) as Input Features and `railne` (node) as Near Features. In the updated attribute table for `City4`, the field `NEAR_FID` identifies the closest node on the railway network to a city, and another field, `NEAR_DIST`, identifies the distance between them. This step completes measuring the air distance from a county to its nearest node on railroads (i.e., segment S1 in Figure 2.3), and the air distance from a city to its nearest node on railroads (i.e., segment S3 in Figure 2.3).

3. *Identifying unique origin and destination nodes*: In network modeling, both the origin and destination nodes need to be unique. In the attribute table for `CntyNEpt`, we can find many cases of multiple counties corresponding to one `NEAR_FID` code. For example, two counties with `FID = 5` and `FID = 8` have the same `NEAR_FID = 34`. In other words, several nearby counties may share the same nearest node (origin node) on the railroad. In the attribute table for `city4`, each city corresponds to one unique node, and thus requires no further processing. There are four unique destination nodes. The following explains how to identify unique origin nodes.
   On the opened attribute table for `CntyNEpt`, right-click the field `NEAR_FID` > choose Summarize > name the output table `Sum_FID.dbf`, where the field `Cnt_NEAR_F` (frequency count) represents how many counties correspond to each `NEAR_FID` code. Any counties with a frequency count greater than 1 indicate that they share one nearest node. The table `Sum_FID.dbf` has 149 records, implying 149 unique origin nodes.

4. *Defining INFO files for origin and destination nodes*: This step prepares two files to be used next: one contains all origin nodes, and another contains all destination nodes. Both need to be in INFO format prepared in ArcInfo Workstation. The dBase table `Sum_FID.dbf` is used to create

the INFO file for origin nodes. The attribute table `city4.pat` is already
an INFO file,[3] based on which the INFO file for destination nodes will
be created. Both tasks are done in ArcInfo Workstation as follows.

In ArcInfo Workstation, navigate to the project directory (e.g., by typing the
command `w c:\Quant_GIS\proj2`) and type the following commands[4]:

```
Dbaseinfo sum_fid.dbf tmp          /*convert to INFO file "tmp"
Pullitems tmp fm_node near_fid /*extract the item "near_fid"
   to create INFO file "fm_node" for origin nodes
Pullitems city4.pat to_node near_fid
   /*extract the item "near_fid" to create INFO file "to_node" for destina-
   tion nodes
```

The item name `near_fid` in both INFO files `fm_node` and `to_node`
needs to be changed to `railne-id` to match the railroad coverage name.
The item `railne-id` is the unique identification number for each node
in the node attribute table `railne.nat`. This can be done in ArcCatalog:
right-click the table `fm_node` (or `to_node`) > choose Properties from
the context menu > click the Items tab to open the dialog window > click
Edit to change the name of an item. Experienced ArcInfo Workstation
users may change an item's name inside the Workstation environment and
write an AML program to automate the process, including the next step.

5. *Computing distances between nodes through railroads*: The following
    commands in ArcInfo Workstation implement the task:

```
ap                                  /* access the arcplot module
netcover railne railroute    /* set up the route system
centers fm_node                /* define the origin nodes
stops to_node                   /* define the destination nodes
nodedistance centers stops rdist 3000000 network ids
q                                    /*exit
```

The "nodedistance" command computes the distance from each node
defined in `centers` to each node defined in `stops`, uses 3000 km
(or a very large distance value) as the search cutoff distance, and creates
an INFO file `rdist`. The final two arguments are optional: "network" is
the default option (the other two are "Euclidean" and "Manhattan," which
compute Euclidean and Manhattan distances respectively) and the option
"ids" specifies that node IDs are used to identify the origin and destination
nodes (the default option is "noids"). In the INFO file `rdist`, the item
`railne-ida` identifies the origin nodes, the item `railne-idb` identifies the destination nodes, and the item `network` is the network
distances between them. This step completes measuring the network
distances from origin nodes to destination nodes (i.e., segment S2 in
Figure 2.3). There are 149 origin nodes in the table `fm_node` and
4 destination nodes in the table `to_node`, and thus $149 \times 4 = 596$ records
in the network distance file `rdist`, which is less than the 812 records in
the Euclidean distance file `Dist.dbf`.

The next task is to join the three distance segments together: S2 is in the table `rdist`, and S1 and S3 are obtained in step 2 in the updated attribute tables for `CntyNEpt` and `city4`, respectively. However, one cannot attempt to join the attribute table `CntyNEpt` to `rdist` in the hope to obtain a table with distance segments S1 and S2.[5] Recall that one origin node may correspond to multiple counties in `CntyNEpt`, as explained in step 3, and one origin node is associated with four destination nodes in `rdist`. Therefore, the relationship between the two tables `CntyNEpt` and `rdist` would be many to many based on the common key "origin nodes." This creates a challenge for creating a table containing three distance segments. We will utilize the Euclidean distance file `Dist.dbf` to accomplish the task, as shown in the next step. Figure 2.4a to c is designed to help readers understand the process.

6. *Attaching the air distance segments*: In ArcMap, right-click the table `Dist.dbf` > choose Joins and Relates > Remove Join(s). This clears the table `Dist.dbf` by dropping unnecessary fields created from previous joins. Similar to step 4 in Part 1 of the project, use "join" twice: join the attribute table of `CntyNEpt` to `Dist.dbf` (common keys are `FID` and `INPUT_FID`, respectively) and join the attribute table `city4` to `Dist.dbf` (common keys are `FID` and `NEAR_FID`, respectively). Note that both the attribute tables are updated with air distance segments in step 2, which are transferred to the combined table `Dist.dbf`: `CntyNEpt.NEAR_DIST` is the distance between counties and their closest nodes, and `point:NEAR_DIST` is the distance between cities and their closest nodes. Figure 2.4a illustrates this step.

7. *Attaching the network distance segment*: In order to join the network distance table `rdist` to `Dist.dbf`, we need to create a common key `linkid` identifying a unique railroad route from an origin node to a destination node. The field `linkid` is made of both the origin node IDs and destination node IDs.

   Open the INFO table `rdist`, add a field `linkid` (define the type as "long integer"), and compute it as `linkid = 1000*railne-ida + railne-idb`. For example, if railne-ida = 198 and railne-idb = 414, then linkid = 198,414. See the left table in Figure 2.4b. Similarly, add the same field `linkid` to the table `Dist.dbf` and compute it as `Dist.linkid = 1000*CntyNEpt.NEAR_FID+point:NEAR_FID`. See the right table in Figure 2.4b. Finally, use the common key `linkid` to join the table `rdist` to `Dist.dbf`.

8. *Summing up three distance segments*: Add a field `RoadDist` (define the type as "float") to `Dist.dbf` and calculate it as `Dist.RoadDist = (CntyNEpt.NEAR_DIST + point:NEAR_DIST + rdist:network)/1000`. The field `RoadDist` in `Dist.dbf` is the total distance from each county to each major city through the railroad network in kilometers. See Figure 2.4c for the final combined table.

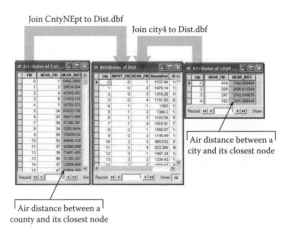

Join CntyNEpt to Dist.dbf

Join city4 to Dist.dbf

Air distance between a
city and its closest node

Air distance between a
county and its closest node

(a)

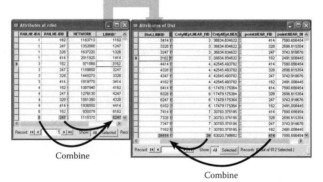

Join rdist to Dist.dbf

Combine

Combine

(b)

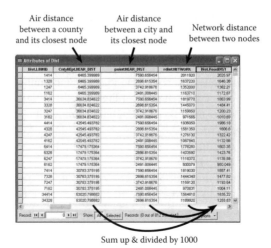

Air distance
between a county
and its closest node

Air distance
between a city and
its closest node

Network distance
between two nodes

Sum up & divided by 1000

(c)

FIGURE 2.4 Table joins in computing travel distances.

2.3.3 PART 3: MEASURING TRAVEL TIME (OPTIONAL)

Setion 2.3.2 has demonstrated how to measure travel distances through a road network. For travel time, the procedures are similar. The following only points out the differences.

In step 1, add an item, `speed`, to the road network attribute table (`railne.aat`) and assign a speed to each road segment; then add another item, `time`, to the same attribute table and calculate it as `time = length/speed`. Pay attention to the units for length and speed, as unit conversions may be needed. For example, if the speed is in kilometers per hour, the formula would be `time = (length/1000) /speed` in hours.

In step 5, prior to the `NODEDISTANCE` command, add a command to define the impedance item: `impedance time`. Now the item `network` in the INFO file `rdist` represents time instead of distance (by default).

In the final step (step 8), it is necessary to make an assumption for the travel speed across the air distances at the two ends though these segments (S1 and S3) are minor. If this speed is assumed to be 50 km/h, the formula for calculating the total travel time (in hours) would be `Dist.roadtime = (CntyNEpt.NEAR_DIST + point:NEAR_DIST) /1000/50 + rdist:network`.

At the end of the project, one may use ArcCatalog to delete unneeded data, but keep the dBase file `Dist.dbf` containing all three distance measures. This distance file will be used in case study 4B.

2.4 SUMMARY

This chapter covers four basic spatial analysis skills:

1. Measuring Euclidean distances
2. Measuring Manhattan distances
3. Measuring network distances
4. Measuring travel time

Both Euclidean and Manhattan distances are fairly easy to obtain in GIS. Computing network distances or travel time requires the road network data and also takes more steps to implement. Several projects in other chapters need to compute Euclidean distances, network distances, or travel time, and thus provide additional practice for developing this basic skill in spatial analysis.

APPENDIX 2: THE VALUED-GRAPH APPROACH TO THE SHORTEST-ROUTE PROBLEM

The *valued graph*, or *L matrix*, provides another way to solve the shortest-route problem (Taaffe et al., 1996, pp. 272–275).

For example, a network is shown in Figure A2.1. The network resembles the highway network in north Ohio, with node 1 for Toledo, 2 for Cleveland, 3 for Cambridge, 4 for Columbus, and 5 for Dayton. We use a matrix L^1 to represent the network, where each cell is the distance on a direct link (one-step link). If there is

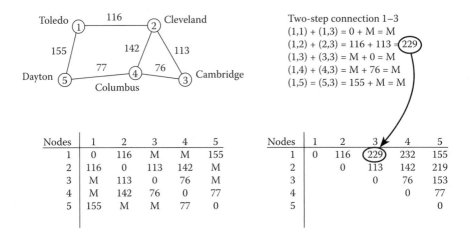

FIGURE A2.1 A valued-graph example.

no direct link between two nodes, the entry is M (a very large number). We enter 0 for all diagonal cells $L^1(i, i)$ because the distance is 0 to connect a node to itself.

The next matrix, L^2, represents two-step connections. All cells in L^1 with values other than M remain unchanged because no distances by two-step connections can be shorter than a one-step (direct) link. We only need to update the cells with the value M. For example, $L^1(1, 3) = M$ needs to be updated. All possible two-step links are examined:

$$L^1(1, 1) + L^1(1, 3) = 0 + M = M$$

$$L^1(1, 2) + L^1(2, 3) = 116 + 113 = 229$$

$$L^1(1, 3) + L^1(3, 3) = M + 0 = M$$

$$L^1(1, 4) + L^1(4, 3) = M + 76 = M$$

$$L^1(1, 5) + L^1(5, 3) = 155 + M = M$$

The cell value $L^2(1, 3)$ is the minimum of all the above links, which is $L^1(1, 2) + L^1(2, 3) = 229$. Note that it records not only the shortest distance from 1 to 3, but also the route (through node 2).

Similarly, other cells are updated, such as $L^2(1, 4) = L^1(1, 5) + L^1(5, 4) = 155 + 77 = 232$, $L^2(2, 5) = L^1(2, 4) + L^1(4, 5) = 142 + 77 = 219$, $L^2(3, 5) = L^1(3, 4) + L^1(4, 5) = 76 + 77 = 153$, and so on. The final matrix L^2 is shown in Figure A2.1 (lower right corner).

By now, all cells in L^2 have values other than M and the shortest-route problem is solved. Otherwise, the process continues until all cells have values other than M. For example, L^3 would be computed as

$$L^3(i, j) = \min\{L^1(i,k) + L^2(k, j), \forall k\}$$

NOTES

1. Spatial data for the counties and cities are extracted from the China county-level GIS data available at http://sedac.ciesin.columbia.edu/china. The railway dataset is provided by Dr. Fengjun Jin at the Institute of Geographical Sciences and Natural Resources Research, Chinese Academy of Sciences.
2. Ideally, nodes should be defined as the railroad stations (stops) in the real world.
3. In ArcInfo Workstation, the attribute table for a polygon or point coverage has a file extension .PAT, which stands for polygon (point) attribute table; the attribute table for a line (arc) coverage has a file extension .AAT; and the attribute table for a node coverage has a file extension .NAT.
4. Texts following "/*" are just a short comment explaining each command.
5. Since each destination node corresponds to a unique city, it would not be a problem to join the attribute table `City4` to `rdist` (based on the common key "destination nodes") in order to obtain a table with distance segments S3 and S2.

3 Spatial Smoothing and Spatial Interpolation

This chapter covers two more generic tasks in GIS-based spatial analysis: spatial smoothing and spatial interpolation. Spatial smoothing and spatial interpolation are closely related and are both useful to visualize spatial patterns and highlight spatial trends. Some methods (e.g., kernel estimation) can be used in either spatial smoothing or interpolation. There are varieties of spatial smoothing and spatial interpolation methods. This chapter only covers those most commonly used.

Conceptually similar to moving averages (e.g., smoothing over a longer time interval), *spatial smoothing* computes the averages using a larger spatial window. Section 3.1 discusses the concepts and methods for spatial smoothing, followed by case study 3A using spatial smoothing methods to examine Tai place-names in southern China in Section 3.2. *Spatial interpolation* uses known values at some locations to estimate unknown values at other locations. Section 3.3 covers point-based spatial interpolation, and Section 3.4 uses case study 3B to illustrate some common point-based interpolation methods. Case study 3B uses the same data and further extends the work in case study 3A. Section 3.5 discusses *area-based spatial interpolation*, which estimates data for one set of (generally larger) areal units with data for a different set of (generally smaller) areal units. Area-based interpolation is useful for data aggregation and integration of data based on different areal units. Section 3.6 presents case study 3C to illustrate two simple area-based interpolation methods. The chapter is concluded with a brief summary in Section 3.7.

3.1 SPATIAL SMOOTHING

Like moving averages that are calculated over a longer time interval (e.g., 5-day moving-average temperatures), *spatial smoothing* computes the value at a location as the average of its nearby locations (defined in a spatial window) to reduce spatial variability. Spatial smoothing is a useful method for many applications. One is to address the *small numbers problem*, which will be explored in detail in Chapter 8. The problem occurs for areas with small populations, where the rates of rare events such as cancer or homicide are unreliable due to random error associated with small numbers. The occurrence of one case can give rise to unusually high rates in some areas, whereas the absence of cases leads to a zero rate in many areas. Another application is for examining spatial patterns of point data by converting discrete point data to a continuous density map, as illustrated in Section 3.2. This section discusses two common spatial smoothing methods (floating catchment area method and kernel estimation), and Appendix 3 introduces the *empirical Bayes estimation*.

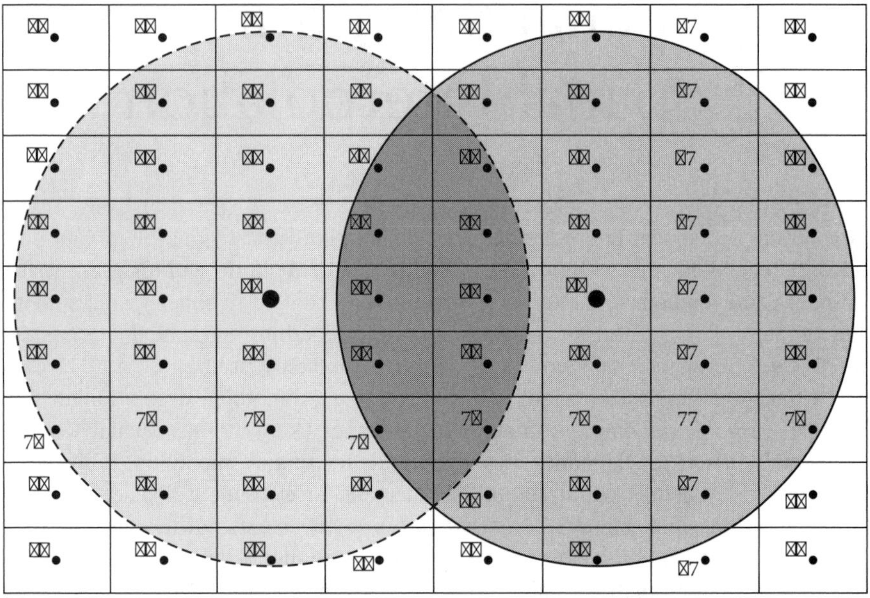

FIGURE 3.1 The FCA method for spatial smoothing.

3.1.1 FLOATING CATCHMENT AREA METHOD

The *floating catchment area* (FCA) *method* draws a circle or square around a location to define a filtering window and uses the average value (or density of events) within the window to represent the value at the location. The window moves across the study area until averages at all locations are obtained. The average values have less variability and are thus spatially smoothed values. The FCA method may be also used for other purposes, such as accessibility measures (see Section 5.2).

Figure 3.1 shows part of a study area with 72 grid-shaped tracts. The circle around tract 53 defines the window containing 33 tracts (a tract is included if its centroid falls within the circle), and therefore the average value of these 33 tracts represents the spatially smoothed value for tract 53. The circle centers around each tract centroid and moves across the whole study area until smoothed values for all tracts are obtained. A circle of the same size around tract 56 includes another set of 33 tracts that defines a new window for tract 56. Note that windows near the borders of a study area do not include as many tracts and cause a lesser degree of smoothing. Such an effect is referred to as *edge effect*.

The choice of window size is very important and should be made carefully. A larger window leads to stronger spatial smoothing, and thus better reveals regional than local patterns; a smaller window generates reverse effects. One needs to experiment with different sizes and choose one with balanced effects.

Implementing the FCA in ArcGIS is demonstrated in case study 3A in detail. We first compute the distances (e.g., Euclidean distances) between all objects, and then distances less than or equal to the threshold distance are extracted.[1] In ArcGIS, we then summarize the extracted distance table by computing average values of

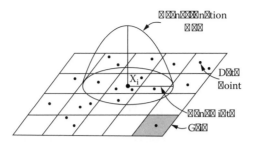

FIGURE 3.2 Kernel estimation.

attributes by origins. Since the table only contains distances within the threshold, only those objects (destinations) within the window are included and form the catchment area in the summarization operation. This eliminates the need of programming that implements iterations of drawing a circle and searching for objects within the circle.

3.1.2 Kernel Estimation

The *kernel estimation* bears some resemblance to the FCA method. Both use a filtering window to define neighboring objects. Within the window, the FCA method does not differentiate far and nearby objects, whereas the kernel estimation weighs nearby objects more than far ones. The method is particularly useful for analyzing and displaying point data. The occurrences of events are shown as a map of scattered (discrete) points, which may be difficult to interpret. The kernel estimation generates a density of the events as a continuous field, and thus highlights the spatial pattern as peaks and valleys. The method may also be used for spatial interpolation.

A kernel function looks like a bump centered at each point x_i and tapering off to 0 over a bandwidth or window. See Figure 3.2 for illustration. The kernel density at point x at the center of a grid cell is estimated to be the sum of bumps within the bandwidth:

$$\hat{f}(x) = \frac{1}{nh^d} \sum_{i=1}^{n} K\left(\frac{x - x_i}{h}\right)$$

where $K(\ )$ is the kernel function, h is the bandwidth, n is the number of points within the bandwidth, and d is the data dimensionality. Silverman (1986, p. 43) provides some common kernel functions. For example, when $d = 2$, a commonly used kernel function is defined as

$$\hat{f}(x) = \frac{1}{nh^2\pi} \sum_{i=1}^{n} [1 - \frac{(x - x_i)^2 + (y - y_i)^2}{h^2}]^2$$

where $(x - x_i)^2 + (y - y_i)^2$ measures the deviation in x-y coordinates between points (x_i, y_i) and (x, y).

Similar to the effect of window size in the FCA method, larger bandwidths tend to highlight regional patterns and smaller bandwidths emphasize local patterns (Fotheringham et al., 2000, p. 46).

ArcGIS has a built-in tool for kernel estimation. To access the tool, make sure that the Spatial Analyst extension is turned on by going to the Tools from the main manual bar and selecting Extensions. Click the Spatial Analyst dropdown arrow > Density > choose Kernel for Density Type in the dialog.

3.2 CASE STUDY 3A: ANALYZING TAI PLACE-NAMES IN SOUTHERN CHINA BY SPATIAL SMOOTHING

This case study examines the distribution pattern of Tai place-names in southern China. The study is part of an ongoing larger project[2] dealing with the historical origins of the Tai in southern China. The Sinification of ethnic minorities, such as the Tai, has been a long and ongoing historical process in China. One indication of historical changes is reflected in geographical place-names over time. Many older Tai names can be recognized because they are named after geographical or other physical features in Tai, such as "rice field," "village," "mouth of a river," "mountain," etc. On the other hand, many other older Tai place-names have been obliterated or modified in the process of Sinification. The objective of the larger project is to reconstruct all the earlier Tai place-names in order to discover the original extent of Tai settlement areas in southern China before the Han pushed south. This case study is chosen to demonstrate the use of GIS technology in historical-linguistic-cultural studies, a field whose scholars are less exposed to it.

We selected Qinzhou Prefecture in Guangxi Autonomous Region, China, as the study area (see the inset in Figure 3.3). Mapping is important for examining spatial patterns, but direct mapping of Tai place-names may not be very informative. Figure 3.3 shows the distribution of Tai and non-Tai place-names, from which we can vaguely see areas with more representations of Tai place-names and others with less. The spatial smoothing techniques help visualize the spatial pattern.

The following datasets are provided in the CD for the project:

1. Point coverage `qztai` for all towns in Qinzhou, with the item `TAI` identifying whether a place-name is Tai (= 1) or non-Tai (= 0).
2. Shapefile `qzcnty` defines the study area of six counties.

3.2.1 PART 1: SPATIAL SMOOTHING BY THE FLOATING CATCHMENT AREA METHOD

We first test the floating catchment area method. Different window sizes are used to help identify an appropriate window size for an adequate degree of smoothing to highlight general trends but not to block local variability. Within the window around each place, the ratio of Tai place-names among all place-names is computed to represent the concentration of Tai place-names *around* that place. In implementation, the key step is to utilize a distance matrix between any two places and extract the places that are within a specified search radius from each place.

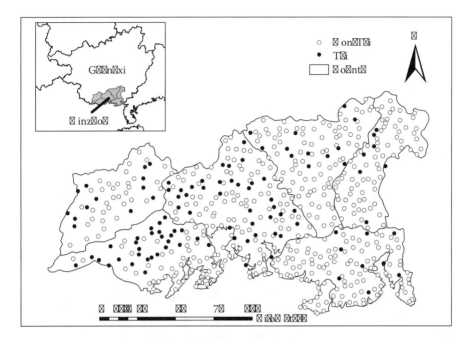

FIGURE 3.3 Tai and non-Tai place-names in Qinzhou.

1. *Computing distance matrix between places*: Refer to Section 2.3.1 for measuring the Euclidean distances. In ArcToolbox, choose Analysis Tools > Proximity > Point Distance. Enter `qztai` (point) as both the Input Features and the Near Features and name the output table `Dist_50km.dbf`. By defining a wide search radius of 50 km, the distance table allows us to experiment with various window sizes ≤ 50 km. In the distance file `Dist_50km.dbf`, the `INPUT_FID` identifies the "from" (origin) place, and the `NEAR_FID` identifies the "to" (destination) place.

2. *Attaching attributes of Tai place-names to distance matrix*: Join the attribute table of `qztai` to the distance table `Dist_50km.dbf` based on the common keys `FID` in `qztai` and `NEAR_FID` in `Dist_50km.dbf`. By doing so, each destination place is identified as either a Tai place or non-Tai place by the field `point:Tai`.

3. *Extracting distance matrix within a window*: For example, we define the window size with a radius of 10 km. Open the table `Dist_50km.dbf` > click the tab Options at the right bottom > Select By Attributes > enter the condition `Dist_50km.DISTANCE <=10000`. For each origin place, only those destination places within 10 km are selected. Click Options > Export, and save the new table as `Dist_10km.dbf`, which keeps only distances of 10 km. Those records with a distance = 0 (i.e., the origin and destination places are the same) indicate that the search circles are centered around these places.

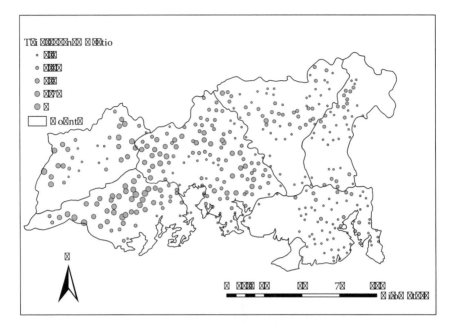

FIGURE 3.4 Tai place-name ratios in Qinzhou by the FCA method.

4. *Calculating Tai place ratios within the window*: On the opened table `Dist_10km.dbf`, right-click the field `INPUT_FID` and choose Summarize > note that `INPUT_FID` appears in the first box (field to summarize), check the field `TAI` (Sum) in the second box (summary statistics), and name the output table `Sum_10km.dbf`. In `Sum_10km.dbf`, the field `Sum_TAI` indicates the number of Tai place-names within a 10-km radius and the field `Count_INPUT_FID` indicates the total number of place-names within the same range. Add a new field `Tairatio` to the table `Sum_10km.dbf` and calculate it as `Tairatio = Sum_TAI / Cnt_INPUT_`. Note that `Cnt_INPUT_` is the abbreviated field name for `Count_INPUT_FID`. This ratio measures the portion of Tai place-names among all places within the window that is centered at each place.

5. *Attaching Tai place-name ratios to the point coverage*: Join the table `Sum_10km.dbf` to the attribute table `qztai` based on the common keys `INPUT_FID` in `Sum_10km.dbf` and `FID` in `qztai`.

6. *Mapping Tai place-name ratios*: Use proportional point symbols to map Tai place-name ratios (each representing the ratio within a 10-km radius around a place) across the study area, as shown in Figure 3.4.
   This completes the FCA method for spatial smoothing, which converts a binary variable `TAI` to a continuous ratio variable `Tairatio`.

7. *Sensitivity analysis*: Experiment with other window sizes, such as 5 and 15 km, and repeat steps 3 to 6. Compare the results with Figure 3.4 to examine the impact of window size. Table 3.1 summarizes the results. As the window size increases, the standard deviation of Tai place-name ratio declines, indicating stronger spatial smoothing.

TABLE 3.1
FCA Spatial Smoothing by Different Window Sizes

Window Size (Radius) (km)	Ratio of Tai Place-Names			
	Min.	Max.	Mean	Std. Dev.
5	0	1.0	0.1868	0.3005
10	0	1.0	0.1886	0.1986
15	0	0.8333	0.1878	0.1642

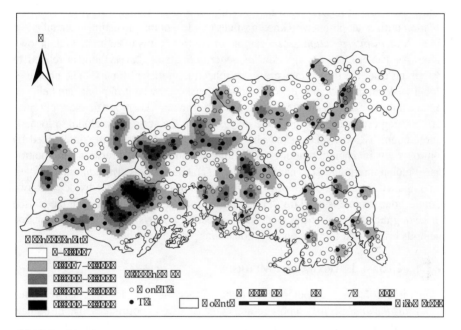

FIGURE 3.5 Kernel density of Tai place-names in Qinzhou.

3.2.2 PART 2: SPATIAL SMOOTHING BY KERNEL ESTIMATION

1. *Execute kernel estimation*: In ArcMap, make sure that the Spatial Analyst extension is turned on: from the Tools menu > choose Extensions > check Spatial Analyst, and from the View menu > choose Toolbars > check Spatial Analyst. Click the Spatial Analyst dropdown arrow > choose Density to activate the dialog window. In the dialog, make sure that qztai (point) is the Input data, select TAI for the Population field, choose kernel as Density type, use 10,000 (meters) for Search radius, square kilometers for Area units, and 1000 (meters) for Output cell size, and name the output raster kernel_10k.

2. *Mapping kernel density*: By default, estimated kernel densities are categorized into nine classes, displayed as different hues. Figure 3.5 is based

on reclassified kernel densities (five classes) with county boundaries as the background.

The kernel density map shows the distribution of Tai place-names as a continuous surface so that patterns like peaks and valleys can be identified. However, the density values simply indicate relative degrees of concentration and cannot be interpreted as a meaningful ratio like `Tairatio` in the FCA method.

3.3 POINT-BASED SPATIAL INTERPOLATION

Point-based spatial interpolation includes global and local methods. A *global interpolation* utilizes all points with known values (control points) to estimate an unknown value. A *local interpolation* uses a sample of control points to estimate an unknown value. As Tobler's (1970) *first law of geography* states, "everything is related to everything else, but near things are more related than distant things." The choice of global vs. local interpolation depends on whether faraway control points are believed to have influence on the unknown values to be estimated. There are no clear-cut rules for choosing one over the other. One may consider the scale from global to local as a continuum. A local method may be chosen if the values are *most* influenced by control points in a neighborhood. A local interpolation also requires less computation than a global interpolation (Chang, 2004, p. 277). One may use *validation* techniques to compare different models. For example, the control points can be divided into two samples: one sample is used for developing the models, and the other sample is used for testing the accuracy of the models. This section surveys two global interpolation methods briefly and focuses on three local interpolation methods.

3.3.1 GLOBAL INTERPOLATION METHODS

Global interpolation methods include trend surface analysis and regression model. *Trend surface analysis* uses a polynomial equation of x-y coordinates to approximate points with known values such as

$$z = f(x,y)$$

where the attribute value z is considered as a function of x and y coordinates (Bailey and Gatrell, 1995). For example, a cubic trend surface model is written as

$$z(x,y) = b_0 + b_1 x + b_2 y + b_3 x^2 + b_4 xy + b_5 y^2 + b_6 x^3 + b_7 x^2 y + b_8 xy^2 + b_9 y^3$$

The equation is usually estimated by an ordinary least squares regression. The estimated equation is then used to project unknown values at other points.

Higher-order models are needed to capture more complex surfaces and yield higher R-square values (goodness of fit) or lower root mean square (RMS) in general.[3] However, a better fit for the control points is not necessarily a better model for estimating unknown values. Validation is needed to compare different models.

If the dependent variable (i.e., the attribute to be estimated) is binary (i.e., 0 and 1), the model is a *logistic trend surface model* that generates a probability surface. A local version of trend surface analysis uses a sample of control points to estimate the unknown value at a location and is referred to as *local polynomial interpolation*.

ArcGIS offers up to 12th-order trend surface model. To access the method, make sure that the Geostatistical Analyst extension is turned on. In ArcMap, click the Geostatistical Analyst dropdown arrow > Explore Data > Trend Analysis.

A *regression model* uses a linear regression to find the equation that models a dependent variable based on several independent variables, and then uses the equation to estimate unknown points (Flowerdew and Green, 1992). Regression models can incorporate both spatial (not limited to *x-y* coordinates) and attribute variables in the models, whereas trend surface analysis only uses *x-y* coordinates as predictors.

3.3.2 LOCAL INTERPOLATION METHODS

The following discusses three popular local interpolators: inverse distance weighted, thin-plate splines, and kriging.

The *inverse distance weighted* (IDW) method estimates an unknown value as the weighted average of its surrounding points, in which the weight is the inverse of distance raised to a power (Chang, 2004, p. 282). Therefore, the IDW enforces Tobler's first law of geography. The IDW is expressed as

$$z_u = \frac{\sum_{i=1}^{s} z_i d_{iu}^{-k}}{\sum_{i=1}^{s} d_{iu}^{-k}}$$

where z_u is the unknown value to be estimated at u, z_i is the attribute value at control point i, d_{iu} is the distance between points i and u, s is the number of control points used in estimation, and k is the power. The higher the power, the stronger (faster) the effect of distance decay is (i.e., nearby points are weighted much higher than remote ones). In other words, distance raised to a higher power implies stronger localized effects.

Thin-plate splines create a surface that predicts the values exactly at all control points and has the least change in slope at all points (Franke, 1982). The surface is expressed as

$$z(x,y) = \sum_{i=1}^{n} A_i d_i^2 \ln d_i + a + bx + cy$$

where x and y are the coordinates of the point to be interpolated, $d_i = \sqrt{(x - x_i)^2 + (y - y_i)^2}$ is the distance from the control point (x_i, y_i), and A_i, a,

b, and c are the $n + 3$ parameters to be estimated. These parameters are estimated by solving a system of $n + 3$ linear equations (see Chapter 11), such as

$$\sum_{i=1}^{n} A_i d_i^2 \ln d_i + a + bx_i + cy_i = z_i;$$

$$\sum_{i=1}^{n} A_i = 0 \; ; \; \sum_{i=1}^{n} A_i x_i = 0 \; ; \; \text{and} \; \sum_{i=1}^{n} A_i y_i = 0$$

Note that the first equation above represents n equations for $i = 1, 2, \ldots, n$, and z_i is the known attribute value at point i.

Thin-plate splines tend to generate steep gradients (overshoots) in data-poor areas, and other methods such as *thin-plate splines with tension*, *regularized splines*, and *regularized splines with tension* have been proposed to mitigate the problem (see Chang, 2004, p. 285). These advanced interpolation methods are grouped as *radial basis functions*.

Kriging (Krige, 1966) models the spatial variation as three components: a spatially correlated component, representing the regionalized variable; a "drift" or structure, representing the trend; and a random error. To measure spatial autocorrelation, kriging uses the measure of *semivariance* (1/2 of variance):

$$\gamma(h) = \frac{1}{2n} \sum_{i=1}^{n} [z(x_i) - z(x_i + h)]^2$$

where n is the number of pairs of the control points that are distance (or spatial lag) h apart and z is the attribute value. In the presence of spatial dependence, $\gamma(h)$ increases as h increases, i.e., nearby objects are more similar than remote ones. A *semivariogram* is a plot showing how the values of $\gamma(h)$ respond to the change of distances h.

Kriging fits the semivariogram with a mathematical function or model and uses it to estimate the semivariance at any given distance, which is then used to compute a set of spatial weights. The effect of using the spatial weights is similar to that in the IDW method, i.e., nearby control points are weighted more than distant ones. For instance, if the spatial weight for each control point i and a point s (to be interpolated) is W_{is}, the interpolated value at s is

$$z_s = \sum_{i=1}^{n_s} W_{is} z_i$$

where n_s is the number of sampled points around the point s, and z_s and z_i are the attribute values at s and i, respectively. Similar to the kernel estimation, kriging can be used to generate a continuous field from point data.

In ArcGIS, all three local interpolation methods are available in the Geostatistical Analyst extension. In ArcMap, click the Geostatistical Analyst dropdown arrow >

Geostatistical Wizard > choose Inverse Distance Weighting, Radial Basis Functions, or Kriging in the Methods frame to invoke the IDW method, various thin-plate spline methods, or kriging methods, respectively. The three local interpolators are also available through Spatial Analyst or 3D Analyst. The Geostatistical Analyst is recommended, as it offers more information and better interface (Chang, 2004, p. 298).

3.4 CASE STUDY 3B: SURFACE MODELING AND MAPPING OF TAI PLACE-NAMES IN SOUTHERN CHINA

This project continues case study 3A by mapping the spatial concentrations of Tai place-names in Qinzhou, China, with various surface modeling techniques. No new datasets are needed for the project. We will utilize the results generated in case study 3A, Part 1, in particular the Tai place-name ratios computed by the FCA method.

3.4.1 PART 1: SURFACE MAPPING BY TREND SURFACE ANALYSIS

1. *Activating the Geostatistical Wizard dialog*: If you have exited from ArcMap without saving the project after case study 3A, repeat step 5 in Section 3.2.1: join the table `Sum_10km.dbf` to the attribute table of `qztai`. In ArcMap, make sure that both Geostatistical Analyst and Spatial Analyst extensions are turned on. Click the Geostatistical Analyst dropdown arrow > choose Geostatistical Wizard to activate the dialog window.

2. *Using trend surface analysis to generate the surface*: In the dialog, choose `qztai` for Input Data, `Sum_10km.Tairatio` for Attribute, and Global Polynomial Interpolation for Methods. In the next dialog, experiment with different powers (e.g., 1, 3, 5, 8, and 10) and review the surfaces and corresponding RMS values. As the power increases, the surface captures more local patterns and yields a smaller RMS. Here we set the power to 10 with RMS = 0.1124, and the surface `Global Polynomial Interpolation Prediction Map` is automatically added to the layers.

3. *Mapping surface for the study area*: Right-click the surface and choose Data > Export to Raster. Name the output raster `trend10`. Click the Spatial Analyst dropdown arrow > choose Options > select `qzcnty` for Analysis mask. Click the Spatial Analyst dropdown arrow again > choose Raster Calculator > double-click `trend10` under Layers to select, and click Evaluate. The resulting raster `Calculation` is the surface for the study area.

   Right-click `Calculation` and choose Properties to improve the map (e.g., under Display, define 30% transparency; under Symbology, redefine the default symbols). Figure 3.6 shows the trend surface map of Tai place-name ratios. The map displays patterns similar to Figure 3.5 by kernel estimation, but highlights more global than local trends. It clearly shows the highest concentration of Tai place-names in the southwest corner stretching toward northeast and other directions. Note that some interpolated values by the trend surface analysis, such as negative values or values higher than 1.00, are unrealistic.

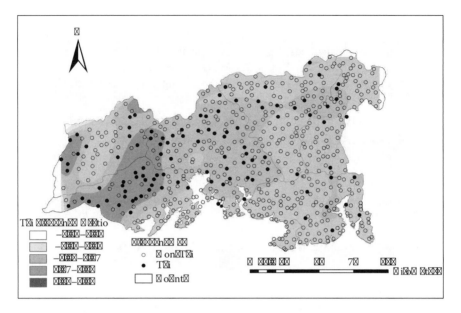

FIGURE 3.6 Interpolated Tai place-name ratios in Qinzhou by trend surface analysis.

4. Optional: *Logistic trend surface analysis*: ArcGIS can also generate a surface directly based on the original binomial (0–1) variable `Tai`. In the dialog window in step 2, choose `point:Tai` for Attribute and others the same as in step 2; repeat the analysis. The result is a probability surface (i.e., the probability of a place being a Tai place-name) generated by logistic trend surface analysis.

3.4.2 PART 2: MAPPING BY LOCAL INTERPOLATION METHODS

1. *Using IDW to map surface*: Similar to step 1 in Part 1, activate the Geostatistical Wizard dialog. Choose `qztai` for Input Data, `Sum_10km.Tairatio` for Attribute, and Inverse Distance Weighting for Methods. Use the default power of 2, 15 neighboring points, and a circular area for selecting control points. Note that RMS = 0.0844. Export the surface to a raster named `idw2`. Similar to step 3 in Part 1, generate a surface for the study area as shown in Figure 3.7. Note that all interpolated values are within the same range as the original, i.e., between 0 and 1. Figure 3.7 shows stronger local patterns than Figure 3.6.

2. *Using thin-plate splines to map surface*: Similarly, in the Geostatistical Wizard dialog window, choose Radial Basis Functions for Methods, and others the same. Use the default method Completely Regularized Spline and other default settings to generate the surface. Name the raster `regspline`. A map based on the surface is slightly different from Figure 3.7 and not shown here.

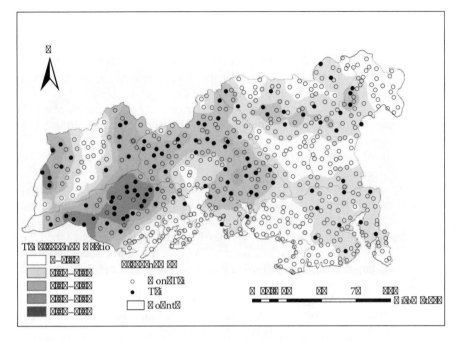

FIGURE 3.7 Interpolated Tai place-name ratios in Qinzhou by the IDW method.

3. *Using kriging to map surface*: Similarly, in the Geostatistical Wizard dialog window, choose Kriging for Methods, and others the same. Use the default method Ordinary Kriging Prediction Map and other default settings to generate the surface. Name the raster `ordkrig`. A similar map is generated and not shown here.

3.5 AREA-BASED SPATIAL INTERPOLATION

Area-based (areal) interpolation is also referred to as *cross-area aggregation*, which transforms data from one system of areal units (source zones) to another (target zones). A point-based method such as kriging or polynomial trend surface analysis can be also used to interpolate a grid surface, and then the values of grid cells are converted to the value for each target zone. In other words, areal units are assumed to be represented by their centroids, and then a point-based method is used to interpolate attributes in areal units as approximation.

Many other methods have been developed for areal interpolation (Goodchild et al., 1993). The simplest and most widely used is *areal weighting interpolator* (Goodchild and Lam, 1980). The method apportions the attribute value from each source zone to target zones according to the areal proportion. The method assumes that the attribute value is evenly distributed within each of the source zones.

More advanced methods may be used to improve interpolation if additional information for the study area is available and utilized. The following discusses a method that is particularly useful for interpolating census data in the U.S. Utilizing the road network information revealed in the U.S. Census Bureau's TIGER files, Xie (1995; also see Batty and Xie, 1994a, 1994b) developed some *network-overlaid algorithms* to project population or other residents-based attributes from one areal unit to another. Residential houses are usually located along the sides of streets or along roads. As a result, the distribution of population is closely related to the street network. Among the three algorithms (network length, network hierarchical weighting, and network house-bearing methods), the *network hierarchical weighting* (NHW) method yields the most promising results.

The critical component of the NHW method is a series of GIS overlaying operations. We take one example to illustrate the method. Researchers on urban issues often use the Census Transportation Planning Package (CTPP) data[4] to analyze land use and transportation issues. For the 1990 CTPP Urban Element data, most regions are aggregated at the traffic analysis zone (TAZ) level.[5] For various reasons (e.g., merging with other census data), it may be desirable to interpolate the CTPP data from TAZs to census tracts. In this case, TAZs are the source zones and census tracts are the target zones. The following five steps implement the task:

1. Overlay the TAZ layer with the census tract layer to create an intersected TAZ–tract (polygon) layer, and overlay the TAZ–tract layer with the road network layer to create a control–net (line) layer.
2. Construct a weight matrix for different road categories, as population or business densities vary along various road classes.
3. Overlay the TAZ layer with the network layer, compute the lengths of various roads and then the weighted length within each TAZ, and allocate population or other attributes to the network.
4. Attach the result from step 3 (population or other attributes) to the control–net layer, and sum up the attributes within each polygon based on the TAZ–tract layer.
5. Sum up attributes by census tracts to get the interpolated attributes within each census tract.

3.6 CASE STUDY 3C: AGGREGATING DATA FROM CENSUS TRACTS TO NEIGHBORHOODS AND SCHOOL DISTRICTS IN CLEVELAND, OHIO

This case study demonstrates two commonly used methods for area-based data aggregation. The first one is not an interpolation method per se, but simply aggregates data from one areal unit to another when multiple source zones are fully included in each target zone (or assumed to be so for approximation). The second method is the areal weighting interpolator. Part 1 illustrates the first method by aggregating census tracts to neighborhoods in Cleveland, as each neighborhood contains multiple complete census tracts. Part 2 illustrates the second method by aggregating census tracts to school districts in Cuyahoga County, where the boundaries of school districts do not match those of census tracts. We use population as an exemplary attribute to demonstrate the interpolation methods.

Data needed for the project are provided in the CD:

1. Shapefile `clevspa2k` for the 36 neighborhoods (statistical planning areas, or SPAs) in Cleveland
2. Shapefile `tgr39035uni` for unified school districts in Cuyahoga County
3. Shapefiles `cuyautm` and `cuya_pt` for census tracts in Cuyahoga County and corresponding centroids

The school district file is downloaded from the Environmental Systems Research Institute, Inc. (ESRI) data website (see case study 1A, step 1). Datasets 3 are generated from case study 1A and 1B.

3.6.1 PART 1: SIMPLE AGGREGATION FROM CENSUS TRACTS TO NEIGHBORHOODS IN THE CITY OF CLEVELAND

Simple aggregation uses a spatial join to transfer data from source zone to target zone, and the implementation is straightforward.

In ArcMap, open the attribute table of `cuyautm`; note that it only contains the field `popuden`, not population. Since attribute joins are temporary, the added fields `area` and `popuden` are retained, but fields from the joined table are lost after exiting the project (case study 1A). Retrieve the field `POP2000` either by joining the table `tgr39035trt00.dbf` again or by adding a field `POP2000` and calculating it as `POP2000 = area * popuden / 1000000`. Repeat the same on the centroid shapefile `cuya_pt`.

Right-click the destination layer `clevspa2k` and choose Joins and Relates > Join. In the dialog, choose the option "Join data from another layer based on spatial location," select `cuya_pt` as the source layer and the summarized join option (which states "Each polygon will be given a summary of the numerical attributes of the points that fall inside it ..."), check the box next to Sum, and name the output shapefile `spa_pop`. Selecting `cuyautm` as the source layer may also work as a polygon-to-polygon join. The safe choice is the point-to-polygon join used here. The field `sum_pop2000` in the attribute table of `spa_pop` is the aggregated population at the neighborhood level (other fields may be deleted).

3.6.2 PART 2: AREAL WEIGHTING AGGREGATION FROM CENSUS TRACTS TO SCHOOL DISTRICTS IN CUYAHOGA COUNTY

1. *Preparing the school district layer*: In ArcToolbox, use Data Management Tools > Projections and Transformations > Feature > Project to transform the shapefile `tgr39035uni` in the geographic coordinate system to a shapefile `cuyauni` in the Universal Transverse Mercator (UTM) coordinate system by importing the projection file defined in `cuyautm`. Add a new variable `area` to the attribute table of `cuyauni` and update it (see Section 1.2, step 3).

FIGURE 3.8 Areal weighting interpolation from census tracts to school districts.

2. *Overlaying census tract and school district layers*: In ArcToolbox, use Analysis Tools > Overlay > Intersect > select `cuyautm` first for Input Features, and then add `cuyauni` also as Input Features, and name the output layer `tmp_int`. In the attribute table of `tmp_int`, the field `area` is the area size of each census tract from `cuyautm`, and the field `area_1` is the area size of each school district from `cuyauni`. Figure 3.8 uses the lower-right corner of the study area to illustrate the intersecting

process. Note that tract 184104 is split into two pieces: polygon A belongs to school district 10016, and B belongs to school district 04660.

3. *Apportioning the attribute to area size*: Add a variable `area_2` to the attribute table of `tmp_int` and update it as the area size of the newly intersected shapefile. Add another field, `popu_est`, to the attribute table of `tmp_int` and compute it as `popu_est = pop2000*area_2/area`. This is the interpolated population for each polygon in the intersected layer by the areal weighting method.[6] For example, the population in polygon A is equal to the population in tract 184104 multiplied by the area size of polygon A divided by the area size of tract 184104, i.e., 1468*1297600/16810900 = 113.

4. *Aggregating data to school districts*: On the attribute table of `tmp_int`, right-click the field `unified` (school district codes) and choose Summarize > select `popu_est` (sum) in the second box, and name the output table `uni_pop.dbf`. The interpolated population for school districts is contained in the field `sum_popu_e` in `uni_pop.dbf`, which may be joined to the layer `cuyauni` for mapping or other purposes.

3.7 SUMMARY

Skills learned in this chapter include:

1. Implementing the FCA method for spatial smoothing
2. Kernel density estimation for mapping point data
3. Trend surface analysis (including logistic trend surface analysis)
4. Local interpolation methods, such as inverse distance weighting, thin-plate splines, and kriging
5. Simple aggregation if multiple source zones are wholly contained in each target zone
6. Areal weighting aggregation if boundaries of source zones do not exactly match those of target zones

Spatial smoothing and spatial interpolation are often used for mapping spatial patterns, like case study 3A and B on Tai place-names in southern China. The techniques are useful in many point-based applications. For example, in case study 4A on defining trade areas for professional sports teams, a simple spatial interpolation method is used to generate a surface map showing the probabilities of residents choosing one club over the other (see Figure 4.4). However, surface mapping is merely descriptive. Identified spatial patterns such as concentrations or lack of concentrations can be arbitrary. Where are concentrations statistically significant instead of random? The answer relies on rigorous statistical analysis, for example, spatial cluster analysis — a topic to be covered in Chapter 9 (case study 9A uses the same dataset to identify spatial clusters of Tai place-names).

Area-based spatial interpolation is often used to convert data from different sources to one areal unit for an integrated analysis. It is also used to convert data from a finer to a coarser resolution for examining the modifiable areal unit problem

(MAUP). For example, in case study 6 on urban density functions, the technique is used to aggregate data from census tracts to townships so that functions based on different areal units can be compared.

APPENDIX 3: EMPIRICAL BAYES (EB) ESTIMATION FOR SPATIAL SMOOTHING

Empirical Bayes (EB) *estimation* is another commonly used method for adjusting or smoothing variables (particularly rates) across areas (e.g., Clayton and Kaldor, 1987; Cressie, 1992). Based on the fact that the joint probability of two events is the product of one event and the probability of the second event conditionally upon the first event, Bayesian inference may be expressed as the inclusion of prior information or belief about a dataset in estimating the probability distribution of the data (Langford, 1994, p. 143), i.e.,

likelihood function × prior belief = posterior belief

Using a disease risk as an example, the likelihood function can be said to be the Poisson distributed numbers of observed cases across a study area. The prior belief is on the distribution of relative risks (rates) conditional on the distribution of observed cases: for example, relative risks in areas of larger population size are likely to be more reliable than those in areas of smaller population size. In summary, (1) the mean rate in the study area is assumed to be reliable and unbiased; (2) rates for large population are adjusted less than rates for small population; and (3) rates follow a known probability distribution.

Assume that a common distribution, gamma, is used to describe the prior distribution of rates. The gamma distribution has two parameters, the shape parameter α and the scale parameter v, with the mean = v/α and the variance = v/α^2. The two parameters α and v can be estimated by a mixed maximum likelihood and moments procedure discussed in Marshall (1991). For an area i with population P_i and k_i cases of disease, the crude incidence rate for the area is k_i/P_i. It can be shown that the posterior expected rate or empirical Bayes estimate is

$$E_i = \frac{k_i + v}{P_i + \alpha}$$

If area i has a small population size, the values of k_i and P_i are small relative to v and α, and the EB estimate E_i will be "shrunken" toward the overall mean v/α. Conversely, if area i has a large population size, the values of k_i and P_i are large relative to v and α, and the EB estimate E_i will be very close to the crude rate k_i/P_i. Compared to the crude rate k_i/P_i, the EB estimate E_i is smoothed by the inclusion of v and α.

EB estimation can be applied to a whole study area where all rates are smoothed toward the overall rate. This is referred to as *global empirical Bayes smoothing*. It can also be applied locally by defining a neighborhood around each area and smoothing the rate toward its neighborhood rate. The process is referred to as *regionalized empirical Bayes smoothing*. A neighborhood for an area can be defined as all its contiguous areas plus itself. Contiguity may be defined as rook contiguity or queen contiguity (see Section 1.4), first-order or second-order contiguity, and so on.

GeoDa, a free package developed by Luc Anselin and his colleagues (http://sal.agecon.uiuc.edu/geoda_main.php), can be used to implement the EB estimation for spatial smoothing. The tools are available in GeoDa 0.9.5-i by choosing `Map > Smooth > Empirical Bayes` (or `Spatial Empirical Bayes`). The `Empirical Bayes` procedure smoothes the rates toward the overall mean in the whole study area, and thus is global EB smoothing. The `Spatial Empirical Bayes` procedure smoothes the rates toward a spatial window around an area (defined as the area and its neighboring areas based on a spatial weights file), and thus is local EB smoothing.

NOTES

1. One may use the threshold distance to set the search radius in distance computation and directly obtain the distances within the threshold. However, a table for distances between all objects gives us the flexibility of experimenting with various window sizes.
2. Collaborators of the project also include John Hartmann and Wei Luo of Northern Illinois University and Jerold Edmondson of the University of Texas at Arlington.
3. RMS is measured as $RMS = \sqrt{\sum_{i=1}^{n} (z_{i,obs} - z_{i,est})^2 / n}$.
4. For the 1990 CTPP, visit http://www.bts.gov/publications/census_transportation_planning_package_1990. For the 2000 CTPP, visit http://www.fhwa.dot.gov/ctpp/.
5. For example, among the 303 CTPP Urban Element regions in 1990, 265 regions are summaries for TAZs, 13 for census tracts, and 25 for block groups (based on the file *Regncode.asc* distributed by the Bureau of Transportation Statistics and summarized by the author).
6. For validation, add a field `popu_valid` and calculate it as `popu_valid = popuden*area_2 /1000000`, which should be identical to `popu_est`. The areal weighting method assumes that population is distributed uniformly within each census tract, and thus a polygon in the intersected layer resumes the population density of the tract, of which the polygon is a component.

Part II

Basic Quantitative Methods and Applications

4 GIS-Based Trade Area Analysis and Applications in Business Geography and Regional Planning

"No matter how good its offering, merchandising, or customer service, every retail company still has to contend with three critical elements of success: location, location, and location" (Taneja, 1999, p. 136). Trade area analysis is a common and important task in the site selection of a retail store. A *trade area* is simply "the geographic area from which the store draws most of its customers and within which market penetration is highest" (Ghosh and McLafferty, 1987, p. 62). For a new store, the study of proposed trading areas reveals market opportunities with existing competitors (including those in the same chain or franchise) and helps decide on the most desirable location. For an existing store, it can be used to project market potentials and evaluate the performance. In addition, trade area analysis provides many other benefits for a retailer: determining the focus areas for promotional activities, highlighting geographic weakness in its customer base, projecting future growth, and others (Berman and Evans, 2001, pp. 293–294).

There are several methods for delineating trade areas: the analog method, the proximal area method, and the gravity models. The analog method is non-geographic, and more recently is often implemented by regression analysis. The proximal area method and the gravity models are geographic approaches and can benefit from GIS technologies. The analog and proximal area methods are fairly simple and are discussed in Section 4.1. The gravity models are the focus of this chapter and are covered in detail in Section 4.2. Because of this book's emphasis on GIS applications, two case studies are presented in Sections 4.3 and 4.4 to illustrate how the two geographic methods (the proximal area method and the gravity models) are implemented in GIS. Case study 4A draws from traditional business geography, but with a fresh angle: instead of the typical retail store analysis, it analyzes the fan bases for two professional baseball teams in Chicago. Case study 4B demonstrates how the techniques of trade area analysis are used beyond retail studies. In this case, the methods are used in delineating hinterlands (influential areas) for major cities in northeast China. Delineation of hinterlands is an important task for regional planning. The chapter is concluded with some remarks in Section 4.5.

4.1 BASIC METHODS FOR TRADE AREA ANALYSIS

4.1.1 ANALOG METHOD AND REGRESSION MODEL

The *analog method*, developed by Applebaum (1966, 1968), is considered the first systematic retail forecasting model founded on empirical data. The model uses an existing store or several stores as analogs to forecast sales in a proposed similar or analogous facility. Applebaum's original analog method did not use regression analysis. The method uses customer surveys to collect data of sample customers in the analogous stores: their geographic origins, demographic characteristics, and spending habits. The data are then used to determine the levels of market penetration (e.g., number of customers, population, and average spending per capita) at various distances. The result is used to predict future sales in a store located in similar environments. Although the data may be used to plot market penetrations at various distances from a store, the major objective of the analog method is to forecast sales but not to define trade areas geographically. The analog method is easy to implement, but has some major weaknesses. The selection of analog stores requires subjective judgment (Applebaum, 1966, p. 134), and many situational and site characteristics that affect a store's performance are not considered.

A more rigorous approach to advance the classical analog method is the usage of *regression models* to account for a wide array of factors that influence a store's performance (Rogers and Green, 1978). A regression model can be written as

$$Y = b_0 + b_1 x_1 + b_2 x_2 + ... + b_n x_n$$

where Y represents a store's sales or profits, x's are explanatory variables, and b's are the regression coefficients to be estimated.

The selection of explanatory variables depends on the type of retail outlets. For example, the analysis on retail banks by Olsen and Lord (1979) included variables measuring trade area characteristics (purchasing power, median household income, homeownership), variables measuring site attractiveness (employment level, retail square footage), and variables measuring level of competition (number of competing banks' branches, trade area overlap with branch of same bank). Even for the same type of retail stores, regression models can be improved by grouping the stores into different categories and running a model on each category. For example, Davies (1973) classified clothing outlets into two categories (corner-site stores and intermediate-site stores) and found significant differences in the variables affecting sales. For corner-site stores, the top five explanatory variables are floor area, store accessibility, number of branches, urban growth rate, and distance to nearest car park. For intermediate-site stores, the top five explanatory variables are total urban retail expenditure, store accessibility, selling area, floor area, and number of branches.

4.1.2 PROXIMAL AREA METHOD

A simple geographic approach for defining trade areas is the *proximal area method*, which assumes that consumers choose the nearest store among similar outlets (Ghosh and McLafferty, 1987, p. 65). This assumption is also found in the classical central place theory (Christaller, 1966; Lösch, 1954). The proximal area method implies that customers only consider travel distance (or travel time as an extension) in their shopping choice, and thus the trade area is simply made of consumers that are closer to the store than any other. Once the proximal area is defined, sales can be forecasted by analyzing the demographic characteristics within the area and surveying their spending habits.

The proximal area method can be implemented in GIS by two ways. The first approach is *consumers based*. It begins with a consumer location and searches for the nearest store among all store locations. The process continues until all consumer locations are covered. At the end, consumers that share the same nearest store constitute the proximal area for that store. In ArcGIS, it is implemented by utilizing the `near` tool in ArcToolbox. The tool is available by invoking Analysis Tools > Proximity > Near.

The second approach is *stores based*. It constructs Thiessen polygons from the store locations, and the polygon around each store defines the proximal area for that store. The layer of Thiessen polygons may then be overlaid with that of consumers (e.g., a census tract layer with population information) to identify demographic structures within each proximal area.[1] In ArcGIS, Thiessen polygons can be generated from a point layer of store locations in *ArcInfo coverage format* by choosing Coverage Tools > Analysis > Proximity > Thiessen. For example, Figure 4.1a to c show how the Thiessen polygons are constructed from five points. First, five points are scattered in the study area as shown in Figure 4.1a. Second, in Figure 4.1b, lines are drawn to connect points that are near each other, and lines are drawn perpendicular to the connection lines at their midpoints. Finally, in Figure 4.1c, the Thiessen polygons are formed by the perpendicular lines.

The proximal area method can be easily extended to use network distance or travel time instead of Euclidean distance. The process implemented in both case studies 4A and 4B follows closely the consumers-based approach. The first step is to generate a distance (time) matrix, containing the travel distance (time) between each consumer location and each store (see Chapter 2). The second step is to identify the store within the shortest travel distance (time) from each consumer location. Finally, the information is joined to the spatial layer of consumers for mapping and further analysis.

4.2 GRAVITY MODELS FOR DELINEATING TRADE AREAS

4.2.1 REILLY'S LAW

The proximal area method only considers distance (or time) in defining trade areas. However, consumers may bypass the closest store to patronize stores with better prices, better goods, larger assortments, or a better image. A store in proximity to other shopping and service opportunities may also attract customers farther than an isolated store because of multipurpose shopping behavior. Methods based on the gravity model consider two factors: distances (or time) from and attractions of stores. Reilly's law of retail gravitation applies the concept of the gravity model to delineating trade areas between two stores (Reilly, 1931). The original Reilly's law was used to define trading areas between two cities.

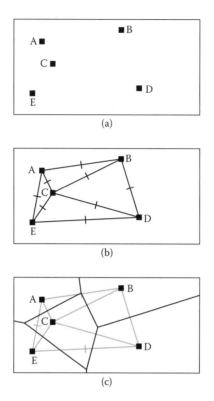

FIGURE 4.1 Constructing Thiessen polygons for five points.

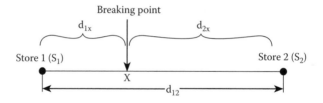

FIGURE 4.2 Breaking point by Reilly's law between two stores.

Consider two stores, stores 1 and 2, that are at a distance of d_{12} from each other (see Figure 4.2). Assume that the attractions for stores 1 and 2 are measured as S_1 and S_2 (e.g., in square footage of the stores' selling areas) respectively. The question is to identify the *breaking point* (BP) that separates trade areas of the two stores. The BP is d_{1x} from store 1 and d_{2x} from store 2, i.e.,

$$d_{1x} + d_{2x} = d_{12} \tag{4.1}$$

By the notion of the *gravity model*, the retail gravitation by a store is in direct proportion to its attraction and in reverse proportion to the square of distance.

Consumers at the BP are indifferent in choosing either store, and thus the gravitation by store 1 is equal to that by store 2, such as

$$S_1 / d_{1x}^{\ 2} = S_2 / d_{2x}^{\ 2} \tag{4.2}$$

Using Equation 4.1, we obtain $d_{1x} = d_{12} - d_{2x}$. Substituting it into Equation 4.2 and solving for d_{1x} yields

$$d_{1x} = d_{12} / (1 + \sqrt{S_2 / S_1}) \tag{4.3}$$

Similarly,

$$d_{2x} = d_{12} / (1 + \sqrt{S_1 / S_2}) \tag{4.4}$$

Equations 4.3 and 4.4 define the boundary between two stores' trading areas and are commonly referred to as Reilly's law.

4.2.2 Huff Model

Reilly's law only defines trade areas between two stores. A more general gravity based method is the *Huff model*, which defines trade areas of multiple stores (Huff, 1963). The model's widespread use and longevity "can be attributed to its comprehensibility, relative ease of use, and its applicability to a wide range of problems" (Huff, 2003, p. 34). The behavioral foundation for the Huff model may be drawn similar to that of the multichoice logistic model. The probability that someone chooses a particular store among a set of alternatives is proportional to the perceived utility of each alternative. That is,

$$P_{ij} = U_j / \sum_{k=1}^{n} U_k \tag{4.5}$$

where P_{ij} is the probability of an individual i selecting a store j, U_j and U_k are the utilities choosing the stores j and k, respectively, and k are the alternatives available ($k = 1, 2, ..., n$).

In practice, the utility of a store is measured as a *gravity kernel*. Like in Equation 4.2, the gravity kernel is positively related to a store's attraction (e.g., its size in square footage) and inversely related to the distance between the store and a consumer's residence. That is,

$$P_{ij} = S_j d_{ij}^{-\beta} / \sum_{k=1}^{n} (S_k d_{ik}^{-\beta}) \tag{4.6}$$

where S is a store's size, d is the distance, $\beta > 0$ is the *distance friction coefficient*, and other notations are the same as in Equation 4.5. Note that the gravity kernel in Equation 4.6 is a more general form than in Equation 4.2, where the distance friction coefficient β is assumed to be 2. The term $S_j d_{ij}^{-\beta}$ is also referred to as *potential*, measuring the impact of a store j on a demand location at i.

Using the gravity kernel to measure utility may be purely a choice of empirical convenience. However, the gravity models (also referred to as *spatial interaction models*) can be derived from individual utility maximization (Niedercorn and Bechdolt, 1969; Colwell, 1982), and thus have its economic foundation (see Appendix 4). Wilson (1967, 1975) also provided a theoretical base for the gravity model by an entropy maximization approach. Wilson's work also led to the discovery of a family of gravity models: a *production-constrained model*, an *attraction-constrained model*, and a *production–attraction-constrained* or *doubly constrained model* (Wilson, 1974; Fotheringham and O'Kelly, 1989).

Based on Equation 4.6, consumers in an area visit stores with various probabilities, and an area is assigned to the trade area of a store that is visited with the highest probability. In practice, given a customer location i, the denominator in Equation 4.6 is identical for various stores j, and thus the highest value of numerator identifies the store with the highest probability. The numerator $S_j d_{ij}^{-\beta}$ is also known as gravity potential for store j at distance d_{ij}. In other words, one only needs to identify the store with the highest potential for defining the trade area. Implementation in ArcGIS can take full advantage of this property. However, if one desires to show a continuous surface of shopping probabilities of individual stores, Equation 4.6 needs to be fully calibrated. In fact, one major contribution of the Huff model is the suggestion that retail trade areas are continuous, complex, and overlapping, unlike the nonoverlapping geometric areas of central place theory (Berry, 1967).

Implementing the Huff model in ArcGIS utilizes a distance matrix between each store and each consumer location, and probabilities are computed by using Equation 4.6. The result is not simply trade areas with clear boundaries, but a continuous probability surface, based on which the simple trade areas can be certainly defined as areas where residents choose a particular store with the highest probability.

4.2.3 Link between Reilly's Law and Huff Model

Reilly's law may be considered a special case of the Huff model. In Equation 4.6, when the choices are only two stores ($k = 2$), $P_{ij} = 0.5$ at the breaking point. That is to say,

$$S_1 d_{1x}^{-\beta} / (S_1 d_{1x}^{-\beta} + S_2 d_{2x}^{-\beta}) = 0.5$$

Assuming $\beta = 2$, the above equation is the same as Equation 4.2, based on which Reilly's law is derived.

For any β, a general form of Reilly's law is written as

$$d_{1x} = d_{12} / [1 + (S_2 / S_1)^{1/\beta}] \tag{4.7}$$

$$d_{2x} = d_{12} / [1 + (S_1 / S_2)^{1/\beta}] \qquad (4.8)$$

Based on Equation 4.7 or 4.8, if store 1 increases its size faster than store 2 (i.e., S_1 / S_2 increases), d_{1x} increases and d_{2x} decreases, indicating that the breaking point (BP) shifts toward store 2 and the trade area for store 1 expands. The observation is straightforward. It is also interesting to examine the impact of the distance friction coefficient on the trade areas. When β decreases, the movement of BP depends on the store sizes:

1. If $S_1 > S_2$, i.e., $S_2 / S_1 < 1$, $(S_2 / S_1)^{1/\beta}$ decreases, and thus d_{1x} increases and d_{2x} decreases, indicating that a larger store is expanding its trade area.
2. If $S_1 < S_2$, i.e., $S_2 / S_1 > 1$, $(S_2 / S_1)^{1/\beta}$ increases, and thus d_{1x} decreases and d_{2x} increases, indicating that a smaller store is losing its trade area.

That is to say, when the β value decreases over time due to improvements in transportation technologies or road network, travel distance matters to a lesser degree, giving even a stronger edge to larger stores. This explains some of the success of superstores in the new era of retail business.

4.2.4 Extensions to the Huff Model

The original Huff model did not include an exponent associated with the store size. A simple improvement over the Huff model in Equation 4.6 is expressed as

$$P_{ij} = S_j^\alpha d_{ij}^{-\beta} / \sum_{k=1}^{n} (S_k^\alpha d_{ik}^{-\beta}) \qquad (4.9)$$

where the exponent α captures elasticity of store size (e.g., a larger shopping center tends to exert more attraction than its size suggests because of scale economies).

The improved model still only used size to measure attractiveness of a store. Nakanishi and Cooper (1974) proposed a more general form called the *multiplicative competitive interaction* (MCI) *model*. In addition to size and distance, the model accounts for factors such as store image, geographic accessibility, and other store characteristics. The MCI model measures the probability of a consumer at residential area i shopping at a store j, P_{ij}, as

$$P_{ij} = (\prod_{l=1}^{L} A_{lj}^{\alpha_l}) d_{ij}^{-\beta} / \sum_{k \in N_i} [(\prod_{l=1}^{L} A_{lk}^{\alpha_l}) d_{ik}^{-\beta}] \qquad (4.10)$$

where A_{lj} is a measure of the lth ($l = 1, 2, ..., L$) characteristic of store j, N_i is the set of stores considered by consumers at i, and other notations are the same as in Equations 4.6 and 4.9.

If disaggregate data of individual shopping trips, instead of the aggregate data of trips from areas, are available, the *multinomial logit* (MNL) *model* is used to model shopping behavior (e.g., Weisbrod et al., 1984), written as

$$P_{ij} = (\prod_{l=1}^{L} e^{\alpha_{lj}A_{lij}})e^{-\beta_{ij}d_{ij}} / \sum_{k\in N_i}[(\prod_{l=1}^{L} e^{\alpha_{lik}A_{lk}})e^{-\beta_{ik}d_{ik}}] \tag{4.11}$$

Instead of using a power function for the gravity kernel in Equation 4.10, an exponential function is used in Equation 4.11. The model is estimated by multinomial logit regression.

4.2.5 DERIVING THE β VALUE IN THE GRAVITY MODELS

The distance friction coefficient β is a key parameter in the gravity models, and deriving its value is an important task prior to the usage of the Huff model. The value varies over time and also across regions, and thus ideally it needs to be derived from the existing travel pattern in a study area.

The original Huff model in Equation 4.6 corresponds to an earlier version of the gravity model for interzonal linkage, written as

$$T_{ij} = aO_i D_j d_{ij}^{-\beta} \tag{4.12}$$

where T_{ij} is the number of trips between zone i (in this case, a residential area) and j (in this case, a shopping outlet), O_i is the size of an origin i (in this case, population in a residential area), D_j is the size of a destination j (in this case, a store size), a is a scalar (constant), and d_{ij} and β are the same as in Equation 4.6. Rearranging Equation 4.12 and taking logarithms on both sides yield

$$\ln[T_{ij} / (O_i D_j)] = \ln a - \beta \ln d_{ij} \tag{4.13}$$

That is to say, if the original model without an exponent for store size is used, the value is derived from a simple bivariate regression model shown in Equation 4.13. See Jin et al. (2004) for an example.

Similarly, the improved Huff model in Equation 4.9 corresponds to a gravity model such as

$$T_{ij} = aO_i^{\alpha_1} D_j^{\alpha_2} d_{ij}^{-\beta} \tag{4.14}$$

where α_1 and α_2 are the added exponents for origin O_i and destination D_j. The logarithmic transformation of Equation 4.14 is

$$\ln T_{ij} = \ln a + \alpha_1 \ln O_i + \alpha_2 \ln D_j - \beta \ln d_{ij} \tag{4.15}$$

Equation 4.15 is the multivariate regression model for deriving the β value if the improved Huff model in Equation 4.9 is used.

4.3 CASE STUDY 4A: DEFINING FAN BASES OF CHICAGO CUBS AND WHITE SOX

In Chicago, it is well known that between the two Major League Baseball (MLB) teams the Cubs outdraw the White Sox in fans regardless of their respective winning records. Many factors, such as history, neighborhoods surrounding the ballparks, pubic images of team management, winning records, and others, may attribute to the difference. In this case study, we attempt to investigate the issue from a geographic perspective. For illustrating trade area analysis techniques, only the population surrounding the ballparks is considered. The proximal area method is first used to examine which club has an advantage if fans choose a closer club. For methodology demonstration, we then consider winning percentage as the only factor for measuring attraction of a club,[2] and use the gravity model method to calibrate the probability surface. For simplicity, Euclidean distances are used for measuring proximity in this project (network distances will be used in case study 4B), and the distance friction coefficient is assumed to be 2, i.e., $\beta = 2$.

Data needed for this project include:

1. A polygon coverage `chitrt` for census tracts in the study area
2. A shapefile `tgr17031lka` for roads and streets in Cook County, where the two clubs are located
3. A comma separated value file `cubsoxaddr.csv` containing the addresses of the clubs and their winning records.

The following explains how the above data sets are obtained and processed. The study area is defined as the 10 Illinois counties in the Chicago consolidated metropolitan statistical area (CMSA) (county codes in parentheses): Cook (031), DeKalb (037), DuPage (043), Grundy (063), Kane (089), Kankakee (091), Kendall (093), Lake (097), McHenry (111), and Will (197). See the inset in Figure 4.3 showing the 10 counties in northeastern Illinois. The spatial and corresponding attribute data are downloaded from the Environmental Systems Research Institute, Inc. (ESRI) data website and processed following procedures similar to those discussed in Section 1.2. The census tract layer of each county is downloaded one at a time and then joined with its corresponding 2000 Census data. Finally, the counties are merged together to form `chitrt` by using the tool in ArcToolbox: Data Management > General > Append. For this project, only the population information from the census is retained, and saved as the field `popu`. One may find other demographic variables, such as income, age, and sex, also useful, and use them for more in-depth analysis.

The shapefile `tgr17031lka` for roads and streets in Cook County, where the two clubs are located, is also downloaded from the ESRI site. This layer is used for geocoding the clubs.

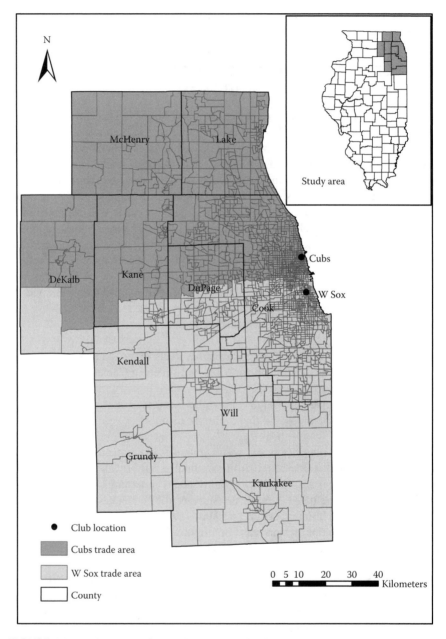

FIGURE 4.3 Proximal areas for the Cubs and White Sox.

Addresses of the two clubs (Chicago Cubs at Wrigley Field, 1060 W. Addison St., Chicago, IL 60613; Chicago White Sox at U.S. Cellular Field, 333 W. 35th St., Chicago, IL 60616) and their winning percentages (0.549 for Cubs and 0.512 for White Sox) in 2003 are found on the Internet and are used to build the file cubsoxaddr.csv with fields club, street, zip, and winrat.

From now on, project instructions will be brief unless a new task is introduced. One may refer to previous case studies for details if necessary. This project introduces a new GIS function, *geocoding* or *address matching*, which enables one to convert a list of addresses into a map of points.

4.3.1 PART 1: DEFINING FAN BASE AREAS BY THE PROXIMAL AREA METHOD

1. *Geocoding the two clubs*: Create a geocoding service in ArcCatalog[3] by the following steps: choose Address Locators > Create New Address Locator > select U.S. Streets with Zone (File) > name the new address locator `mlb`; under Primary table, Reference data, choose `tgr17031lka`; other default values are okay.

   Match addresses in ArcMap by choosing Tools > Geocoding > Geocode Address. Select `mlb` as the address locator, choose `cubsoxaddr.csv` as the address table, and save the result as a shapefile `cubsox_geo`. Project the shapefile to `cubsox_prj` using the projection file defined in the coverage `chitrt` (State Plane Illinois East).

2. *Finding the nearest clubs*: Generate a point layer `chitrtpt` for the centroids of census tracts from the polygon coverage `chitrt` (see Section 1.4, step 1).[4] Use spatial join or the proximity tool in ArcToolbox (Analysis Tools > Proximity > Near) to identify the nearest club from each tract centroid, and attach the result to the polygon coverage `chitrt` for mapping. Figure 4.3 shows the fan base areas for the two clubs defined by the proximal area method.

   If it is desirable to have each trade area shown as an individual polygon (not necessarily for the purpose of this project), one may use ArcToolbox > Data Management Tools > Generalization > Dissolve to group tracts that are assigned to the fan base area of each club.

3. *Summarizing results*: Open the attribute table of `chitrt` and summarize the population (`popu`) by clubs (e.g., `NEAR_FID`). Use Options > Select By Attributes to create subsets of the table that contain tracts within 2 miles (= 3218 m), 5 miles (= 8045 m), 10 miles (= 16,090 m), and 20 miles (= 32,180 m), and summarize the total population near each club. The results are summarized in Table 4.1. It shows a clear advantage for the Cubs, particularly in short-distance ranges. If resident income is considered, the advantage is even stronger for the Cubs.

TABLE 4.1
Fan Bases for Cubs and White Sox by Trade Area Analysis

| Club | By the Proximal Area Method | | | | By Huff Model |
	2 miles	5 miles	10 miles	Study Area	
Cubs	241,297	1,010,673	1,759,721	4,482,460	4,338,884
White Sox	129,396	729,041	1,647,852	3,894,141	4,037,717

4. Optional: *Using the Thiessen polygons to define proximal areas*: In Arc-Toolbox, convert the shapefile `cubsox_prj` to a point coverage `cubsox_pt` using Conversion Tools > To Coverage > Feature Class To Coverage. Use Coverage Tools > Analysis > Proximity > Thiessen to generate a Thiessen polygon coverage `thiess` based on `cubsox_pt`. Use a spatial join (or other overlay tools) to identify census tract centoids that fall within each polygon of `thiess`, and summarize the population for each club. Compare the result to that obtained in step 2. The spatial extent of Thiessen polygons depends on the map extent of the point coverage, and thus may not cover the whole study area.

4.3.2 PART 2: DEFINING FAN BASE AREAS AND MAPPING PROBABILITY SURFACE BY THE HUFF MODEL

1. *Computing distance matrix between clubs and tracts*: Compute the Euclidean distances between the tracts and the clubs in ArcToolbox by choosing Analysis Tools > Proximity > Point Distance (e.g., using `chitrtpt` as Input Feature and `cubsox_prj` as Near Feature; also see Section 2.3, step 2). Name the distance file `dist.dbf`. The distance file has 1902 (number of tracts) × 2 (number of clubs) = 3804 records.

2. *Measuring potential*: Join the attribute table of `cubsox_prj` to `dist.dbf` so that the information of winning records is attached to the distance file. Add a new field `potent` to `dist.dbf`, and calculate it as `potent = 1000000*winrat/(distance/1000)^2`. Note that the values of potential do not have a unit; multiplying it by a constant 1,000,000 is to avoid values being too small. The field `potent` returns the values for the numerator in Equation 4.6.

3. *Calculating probabilities*: On the table `dist.dbf`, sum the field `potent` by census tracts (i.e., `INPUT_FID`) to obtain the dominator term in Equation 4.6 and save the result as `sum_potent.dbf`. Join the table `sum_potent.dbf` back to `dist.dbf`, add a field `prob`, and calculate it as `prob = potent/sum_potent`. The field `prob` returns the probability of residents in each tract choosing a particular club.

4. *Mapping probability surface*: Extract the probabilities of visiting the Cubs (e.g., by selecting the records from `dist.dbf` using the condition `NEAR_FID = 0`) and save the output as `Cubs_Prob.dbf`. Join the table `Cubs_Prob.dbf` to the census tract point layer `chitrtpt` and use the surface modeling techniques in case study 3B to map the probability surface for the Cubs. The result is shown in Figure 4.4. The inset is the zoom-in area near the two clubs, showing the change from one trade area to another along the 0.50 probability line. This case study only considers two clubs. One may repeat the analysis for the White Sox, and the result will be a reverse of Figure 4.4, since probability of visiting the White Sox = 1 − probability of visiting the Cubs.

5. *Defining fan bases by the Huff model*: After the join in step 4, the attribute table of `chitrtpt` has a field `prob`, indicating the probability of

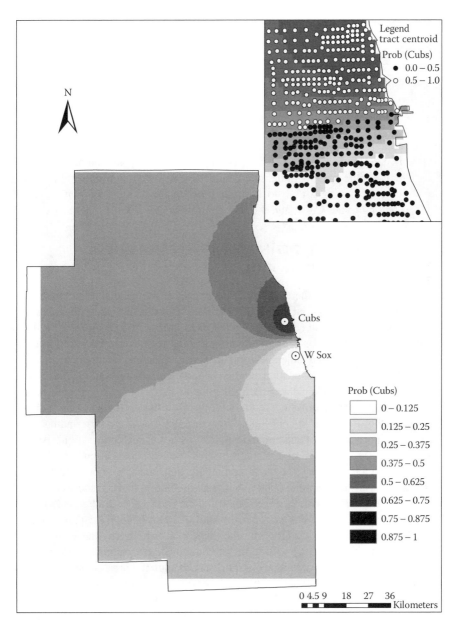

FIGURE 4.4 Probabilities for choosing the Cubs by Huff model.

residents visiting the Cubs. Add a field `cubsfan` to the table and calculate it as `cubsfan = prob * popu`. Summing up the field `cubsfan` yields 4,338,884, which is the projected fan base for the Cubs by the Huff model. The remaining population is projected to be the fan base for the White Sox, i.e., 8,376,601 (total population in the study area) − 4,338,884 = 4,037,717.

4.3.3 Discussion

The proximal area method defines trade areas with definite boundaries. Within a trade area, all residents are assumed to choose one club over the other. The Huff model computes the probabilities of residents choosing each club. Within each tract, a portion of residents chooses one club and the remaining chooses the other. The Huff model seems to produce a more logical result, as real-world fans of different clubs often live in the same area (even in the same household). The model accounts for the impact of each club's attraction though its measurement is usually complex (or problematic, as in this case study). The Huff model may also be used to define the traditional trade areas with definite boundaries by assigning tracts of the highest probabilities of visiting a club to its trade area. In this case, tracts with a `prob` of >0.50 belong to the Cubs, and the remaining tracts are for the White Sox.

4.4   CASE STUDY 4B: DEFINING HINTERLANDS OF MAJOR CITIES IN NORTHEAST CHINA

This section presents another case study that utilizes the techniques of trade area analysis. Instead of traditional applications in retail analysis, this study illustrates how the techniques can be applied to defining the hinterlands of major cities in northeast China. Similar methods for defining urban influential regions can be found in Berry and Lamb (1974), among others. An urban system planning or regional planning project often begins with delineation of urban hinterlands. In Wang (2001a), hinterlands of 17 central cities in China were defined prior to the analysis of regional density functions and growth patterns. Ideally, delineation of urban hinterlands should be based on information of economic connection between cities and their surrounding areas, such as transportation and telecommunication flows or financial transactions. An area is assigned to the hinterland of a city if it has the strongest connection with this city, among other cities. However, data of communication, transportation, and financial flows are often costly or hard to obtain, as is the case for this study. Trade area analysis techniques such as the proximal area method and the Huff model can be used to define hinterlands approximately. For example, the Huff model is built on the gravity model. If residents in an area visit a city with the highest probability (by the Huff model), this implies that the interaction (in terms of communication or transportation flows) between the area and the city is the strongest, and thus the area is assigned to the influence region (hinterland) of the city. Unlike case study 4A, this project uses network distances instead of Euclidean distances. For the reason explained previously (see Section 2.3), distances through the railroads are used to represent the travel distances.

Datasets needed for the project are the same as in case study 2. In addition, the project will use the distance file `Dist.dbf` generated from case study 2 (also provided in the CD for your convenience). Population is used as the attraction measure in the Huff model (i.e., S in Equation 4.6) and is provided in the field `popu_nonfarm` in the point coverage `city4`. We use nonfarm population (according to the 1990 Census), a common index for measuring city sizes in China, to represent population size of the four major cities. See Table 4.2. Among the four major cities, Shenyang, Changchun, and Harbin are the provincial capital cities of

TABLE 4.2
Four Major Cities and Hinterlands in Northeast China

		No. of Counties	
Major City	Nonfarm Population	Proximal Areas	Huff Model
Dalian	1,661,127	7	7
Shenyang	3,054,868	72	72
Changchun	2,192,320	32	24
Harbin	2,990,921	92	100

Liaoning, Jilin, and Heilongjiang, respectively; Dalian is a coastal city that has experienced significant growth after the 1978 economic reform. In the Huff model, we assume $\beta = 2.0$ for convenience.[5]

4.4.1 PART 1: DEFINING PROXIMAL AREAS BY RAILROAD DISTANCES

1. *Extracting distances between counties and their closest cities*: Open the distance file `dist.dbf`, use the tool `Summarize` to identify the minimum railroad distances (i.e., `RoadDist`) by major cities (i.e., `NEAR_FID`), and name the output file `min_rdist.dbf`. The output file contains the fields `INPUT_FID` (identifying county centroid), `Count_INPUT_FID` (= 4 for all counties), and `Minimum_RoadDist` (the distance between a county and its closest city among four major cities), but does not contain any identification information for the corresponding cities.
2. *Identifying the closest cities*: Join the table `min_rdist.dbf` to `dist.dbf`, select the records using the criterion `RoadDist = Minimum_RoadDist`, and export the data to a file `NearCity_id.dbf`. By doing so, a subset of the distance matrix file is created, with 203 records showing each county (identified by `INPUT_FID`) and its closest major city (identified by `NEAR_FID`) by railroads.
3. *Mapping the proximal areas*: Join the table `NearCity_id.dbf` to the county centroid shapefile `CntyNEpt` and then to the county polygon layer `cntyne` for mapping.[6]

Figure 4.5 shows the proximal areas for the four major cities in northeast China. One may also derive the proximal areas based on Euclidean distances and compare the result to Figure 4.5.

4.4.2 PART 2: DEFINING HINTERLANDS BY THE HUFF MODEL

The procedures are similar to those in case study 4A, Part 2.

1. *Measuring potential*: Join the attribute table of `city4` to the distance table `dist.dbf` so that the information for city sizes is attached to the distance table. Add a new field `potent` to `dist.dbf`, and calculate it as `potent = popu_nonfarm/RoadDist^2`.

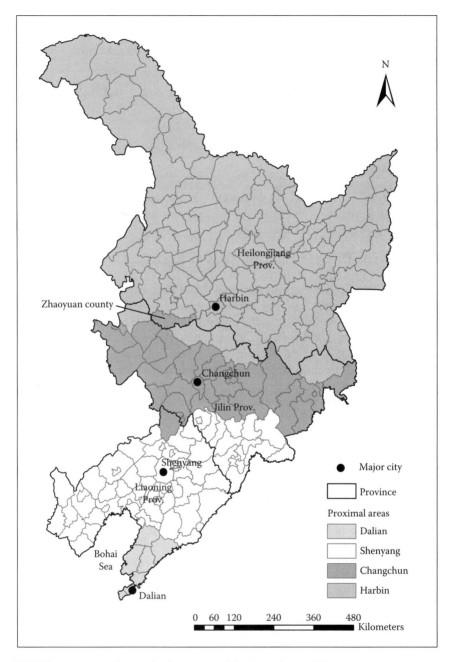

FIGURE 4.5 Proximal areas for four major cities in northeast China.

2. *Identifying cities with the highest potential*: For the purpose of this project, we only need to identify which city (among four major cities) exerts the highest influence (potential) on a county, i.e., the maximum $S_j d_{ij}^{-\beta}$ for

$j = 1, 2, 3$, and 4. For a particular county i, the denominator $\sum_{k=1}^{n}(S_k d_{ik}^{-\beta})$ in Equation 4.6 is the same for any city j; thus, the highest potential implies the highest probability, i.e., $P_{ij} = S_j d_{ij}^{-\beta} / \sum_{k=1}^{n}(S_k d_{ik}^{-\beta})$.

On the table `dist.dbf`, use the tool Summarize to extract the maximum `potent` by counties (i.e., `INPUT_FID`) and save the result as `max_potent.dbf`. Join the table `max_potent.dbf` back to `dist.dbf`, select records with the criterion `dist.potent = max_potent.max_potent`, and export to a table `Maxinfcity.dbf`. The output table `Maxinfcity.dbf` identifies which city has the highest influence (potential) on each county.

3. *Mapping hinterlands of major cities*: Join the table `MaxinfCity.dbf` to the county centroid shapefile `CntyNEpt` and then to the county polygon layer `cntyne` for mapping. Figure 4.6 shows the hinterlands of four major cities in northeast China by the Huff model.

4.4.3 DISCUSSION

Two observations can be made in Figure 4.5. First, a county (Zhaoyuan) in southwest Heilongjiang Province appears closer to Harbin than to Changchun, but is in the proximal area of Changchun based on the railroad distances. This becomes evident by examining the railroad network in Figure 2.2. Second, some counties at the southwest corner of the study area are closer to Dalin than to Shenyang in terms of Euclidean distances but not by railroads. If the proximal areas were based on Euclidean distances, these counties would be assigned to the hinterland of Dalian. Historically, these counties have closer economic ties with Shenyang, and thus belong to its hinterland. This clearly demonstrates the advantage of using network distances for measuring proximity. However, an important developing trend is the rising role of waterway transportation across the Bohai Sea, and this may enhance the economic linkage between these counties and Dalin and change the current boundaries of hinterlands based on the railroads. Figure 4.6 is based on the Huff model accounting for the impact of city sizes. Compared to Figure 4.5, the hinterlands of Shenyang and Dalian are the same as those defined by the proximal area method. However, Figure 4.6 shows an expanded hinterland of Harbin to include some counties closer to Changchun, reflecting the impact of a larger population size of Harbin.

4.5 CONCLUDING REMARKS

While the concepts of proximal area method and the Huff model are straightforward, their successful implementation relies on adequate measurements of variables, which remains one of the most challenging tasks in trade area analysis.

First, both methods use distance or time. The proximal area method is based on the commonly known *least-effect principle* in geography (Zipf, 1949). As shown in case study 4B, road network distance or travel time is generally a better measure

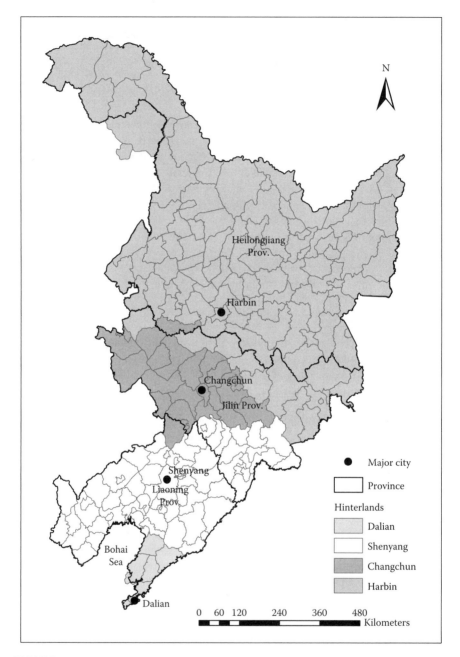

FIGURE 4.6 Hinterlands for four major cities in northeast China by Huff model.

than straight-line (Euclidean) distance. However, network distance or travel time may not be the best measure for travel impedance. Travel cost, convenience, comfort, or safety may also be important. Research indicates that people of various socio-economic or demographic characteristics perceive the same distance differently, i.e., a difference between *cognitive* and *physical distances* (Cadwallader, 1975). Defining network distance or travel time also depends on the particular transportation mode. Case study 4B uses railway, as it is currently the dominant mode for both passenger and freight transportation in China. Similar to the U.S. experience, both air and highway transportations are gaining more ground in China, and waterway can be important in some areas. This makes distance or time measurement more than a routine task. Accounting for interactions by telecommunication, Internet, and other modern technologies adds further complexity to the issue.

Second, in addition to distance or time, the Huff model has two more variables: attraction and travel friction coefficient (S and β in Equation 4.6). Attraction is measured by winning percentage in case study 4A and population size in case study 4B. Both are oversimplification. More advanced methods may be employed to consider more factors in measuring the attraction (e.g., the multiplicative competitive interaction or MCI model). The travel friction coefficient β is also difficult to define, as it varies across time and space, between transportation modes, and is dependent on type of commodities, etc. For additional practice of trade area analysis methods, one may conduct the trade area analysis of chain stores in a familiar study area. Store addresses can be found on the Internet or in other sources (yellow pages, store directories) and geocoded by the procedure discussed in case study 4A. Population census data can be used to measure customer bases. A trade area analysis of the chain stores may be used to project market potentials and evaluate the performance of individual stores.

APPENDIX 4: ECONOMIC FOUNDATION OF THE GRAVITY MODEL

The gravity model is often criticized, particularly by economists, for its lack of foundation in individual behavior. This appendix follows the work of Colwell (1982) in an attempt to provide a theoretical base for the gravity model. For a review of other approaches to derive the gravity model, see Fotheringham et al. (2000, pp. 217–234).

Assume a trip utility function in a Cobb–Douglas form, such as

$$u_i = ax^\alpha z^\gamma t_{ij}^{\tau_{ij}} \tag{A4.1}$$

where u_i is the utility of an individual at location i, x is a composite commodity (i.e., all other goods), z is leisure time, t_{ij} is the number of trips taken by an individual at i to j, $\tau_{ij} = \beta P_j^\phi / P_i^\xi$ is the trip elasticity of utility that is directly related to the destination population P_j and reversely related to the origin population P_i, and α, β, γ, ϕ, and ζ are positive parameters. Colwell (1982, p. 543) justifies the particular way of defining trip elasticity of utility on the ground of central place theory: larger places serve many of the same functions as smaller places, plus higher-order functions not found in smaller places; thus, the elasticity τ_{ij} is larger for trips from the smaller to the larger place than for trips from the larger to the smaller place.

The budget constraint is written as

$$px + rd_{ij}t_{ij} = wW \tag{A4.2}$$

where p is the price of x, r is the unit distance cost for travel, d_{ij} is the distance between point i and j, w is the wage rate, and W is the time worked.

In addition, the time constraint is

$$sx + hd_{ij}t_{ij} + z + W = H \tag{A4.3}$$

where s is the time required per unit of x consumed, h is the travel time per unit of distance, and H is total time.

Combining the two constraints in Equations A4.2 and A4.3 yields

$$(p + ws)x + (rd_{ij} + whd_{ij})t_{ij} + wz = wH \tag{A4.4}$$

Maximizing the utility in Equation A4.1 subject to the constraint in Equation A4.4 yields the following Lagrangian function:

$$L = ax^{\alpha}z^{\gamma}t_{ij}^{\tau_{ij}} - \lambda[(p + ws)x + (rd_{ij} + whd_{ij})t_{ij} + wz - wH]$$

Based on the four first-order conditions, i.e., $\partial L/\partial x = \partial L/\partial z = \partial L/\partial t_{ij} = \partial L/\partial \lambda = 0$, we can solve for t_{ij} by eliminating λ, x, and z:

$$t_{ij} = \frac{wH\tau_{ij}}{(r + wh)d_{ij}(\alpha + \gamma + \tau_{ij})} \tag{A4.5}$$

It is assumed that travel cost per unit of distance r is a function of distance d_{ij}, such as

$$r = r_0 d_{ij}^{\sigma} \tag{A4.6}$$

where $r_0 > 0$ and $\sigma > -1$, so that total travel costs are an increasing function of distance. Therefore, the travel time per unit of distance, h, has a similar function:

$$h = h_0 d_{ij}^{\sigma} \tag{A4.7}$$

so that travel time is proportional to travel cost. For simplicity, assume that the utility function is homogeneous to degree 1, i.e.,

$$\alpha + \gamma + \tau_{ij} = 1 \tag{A4.8}$$

Substituting Equations A4.6, A4.7, and A4.8 into Equation A4.5 and using $\tau_{ij} = \beta P_j^{\varphi} / P_i^{\xi}$, we obtain

$$t_{ij} = \frac{wH\beta P_i^{-\xi} P_j^{\varphi}}{(r_0 + wh_0)d_{ij}^{1+\sigma}} \tag{A4.9}$$

Finally, multiplying Equation A4.9 by the origin population yields the total number of trips from i to j:

$$T_{ij} = P_i t_{ij} = \frac{wH\beta P_i^{1-\xi} P_j^{\varphi}}{(r_0 + wh_0)d_{ij}^{1+\sigma}} \tag{A4.10}$$

which resembles the gravity model in Equation 4.14.

NOTES

1. The coverage of Thiessen polygons is based on the points from which it is generated, and its extent may not cover all consumer locations.
2. Evidently this is an oversimplification. Despite their subpar records for many years, the Cubs have earned the nickname "lovable losers," as one of the most followed clubs in professional sports. However, the record still matters, as tickets to Wrigley Field became harder to get in 2004 after a rare play-off run by the Cubs in 2003. This became more ironic in 2005 when the White Sox earned the best record in the American League and eventually won the World Series.
3. Alternatively, in ArcToolBox, Geocoding Tools > Create Address Locator. However, the interface in ArcCatalog is recommended as it provides more options.
4. One may also use the shapefile `chitrtcent` (population-weighted tract centroids) provided in the CD. Section 5.4.1 discusses how the shapefile is obtained.
5. This is also close to $\beta = 2.1$, a value obtained by Yang (1990) in his study of gravity models for analyzing the interregional passenger flow patterns in China.
6. One may need to export the combined table from the first join, and then join the exported table to the polygon layer so that the fields contained in `NearCity_id.dbf` will not be lost in the second join.

5 GIS-Based Measures of Spatial Accessibility and Application in Examining Health Care Access

Accessibility refers to the relative ease by which the locations of activities, such as work, shopping, recreation, and health care, can be reached from a given location. Accessibility is an important issue for several reasons. Resources or services are scarce, and their efficient delivery requires adequate access by people. The spatial distribution of resources or services is not uniform and needs careful planning and allocation to match demands. Disadvantaged population groups (low-income and minority residents) often suffer from poor access to certain activities or opportunities because of their lack of economic or transportation means. Access can thus become a social justice issue, which calls for careful planning and effective public policies by government agencies.

Accessibility is determined by the distributions of supply and demand and how they are connected in space, and thus is a classic issue for location analysis well suited for GIS to address. This chapter focuses on how spatial accessibility is measured by GIS-based methods. Section 5.1 overviews the issues on accessibility, followed by two GIS-based methods for defining spatial accessibility: the floating catchment area method in Section 5.2 and the gravity-based method in Section 5.3. Section 5.4 illustrates how the two methods are implemented in a case study of measuring access to primary care physicians in the Chicago region. The chapter is concluded with extended discussion and a brief summary.

5.1 ISSUES ON ACCESSIBILITY

Access may be classified according to two dichotomous dimensions (potential vs. revealed, and spatial vs. aspatial) into four categories, such as potential spatial access, potential aspatial access, revealed spatial access, and revealed aspatial access (Khan, 1992). *Revealed accessibility* focuses on actual use of a service, whereas *potential accessibility* signifies the probable utilization of a service. The revealed accessibility may be reflected by frequency or satisfaction level of using a service, and thus be obtained in a survey. Most studies examine potential accessibility, based on which planners and policy analysts evaluate the existing system of service delivery and identify strategies for improvement. *Spatial access* emphasizes the importance of spatial separation between supply and demand as a barrier or a facilitator, whereas *aspatial access* stresses nongeographic barriers or facilitators (Joseph and Phillips, 1984). Aspatial access is related to many demographic and socioeconomic variables. In a study on job access, Wang (2001b) examined how workers' characteristics, such as race, sex, wages, family structure, educational attainment, and homeownership status, affect commuting time and thus job access. In the study on health care access, Wang and Luo (2005) included several categories of aspatial variables: demographics such as age, sex, and ethnicity; socioeconomic status such as population in poverty, female-headed households, homeownership, and income; environment such as residential crowdedness and housing units' lack of basic amenities; linguistic barrier and service awareness such as population without a high school diploma and households linguistically isolated; and transportation mobility such as households without vehicles. Since these variables are often correlated to each other, they may be consolidated into a few factors by using the principal components and factor analysis techniques (see Chapter 7).

This chapter focuses on measuring *potential spatial accessibility*, an issue particularly interesting to geographers and location analysts.

If the capacity of supply is less a concern, one can use simple *supply-oriented accessibility measures* that emphasize the proximity to supply locations. For instance, Brabyn and Gower (2003) used minimum travel distance (time) to the closest service provider to measure accessibility to general medical practitioners in New Zealand. Distance or time from the nearest provider can be obtained using the techniques illustrated in Chapter 2. Hansen (1959) used a simple gravity-based *potential model* to measure accessibility to jobs. The model is written as

$$A_i^H = \sum_{j=1}^{n} S_j d_{ij}^{-\beta},$$ (5.1)

where A_i^H is the accessibility at location i, S_j is the supply capacity at location j, d_{ij} is the distance or travel time between the demand (at location i) and a supply location j, β is the travel friction coefficient, and n is the total number of supply locations. The superscript H in A_i^H denotes the measure based on the Hanson model vs. F for the measure based on the two-step floating catchment area method in Equation 5.2 or G for the measure based on the gravity model in Equation 5.3. The potential model values supplies at all locations, each of which is discounted by a distance term. The model does not account for the demand side. That is to say, the amount of population competing for the supplies is not considered to affect accessibility. The model is the foundation for a more advanced gravity-based method that will be explained in Section 5.3.

In most cases, accessibility measures need to account for both supply and demand because of scarcity of supply. Prior to the widespread use of GIS, the *simple supply–demand ratio method* computed the ratio of supply vs. demand in an area (usually an administrative unit such as township or county) to measure accessibility. For example, Cervero (1989) and Giuliano and Small (1993) measured job accessibility by the ratio of jobs vs. resident workers across subareas (central city and combined suburban townships) and used the ratio to explain intraurban variations of commuting time. In the literature on job access and commuting, the method is commonly referred to as the *jobs–housing balance approach*. The U.S. Department of Health and Human Services (DHHS) uses the population-to-physician ratio within a rational service area (most as large as a whole county or a portion of a county or established neighborhoods and communities) as a basic indicator for defining physician shortage areas[1] (GAO, 1995; Lee, 1991). In the literature on health care access and physician shortage area designation, the method is referred to as the *regional availability measure* (vs. the *regional accessibility measure* based on a gravity model) (Joseph and Phillips, 1984).

The simple supply–demand ratio method has at least two shortcomings. First, it cannot reveal the detailed spatial variations within the areas (usually large). For example, the job–housing balance approach computes the jobs–resident workers ratio and uses it to explain commuting across cities, but cannot explain the variation within a city. Second, it assumes that the boundaries are impermeable; i.e., demand is met by supply only within the areas. For instance, in physician shortage area designation by the DHHS, the population-to-physician ratio is often calculated at the county level, implying that residents do not visit physicians beyond county borders.

The next two sections discuss a two-step floating catchment area (2SFCA) method and a more advanced gravity-based model, respectively. Both methods consider supply and demand and overcome the shortcomings mentioned above.

5.2 THE FLOATING CATCHMENT AREA METHODS

5.2.1 Earlier Versions of Floating Catchment Area Method

Earlier versions of the *floating catchment area* (FCA) *method* are very much like the one discussed in Section 3.1 on spatial smoothing. For example, in Peng (1997), a catchment area is defined as a square around each location of residents, and the jobs–residents ratio within the catchment area measures the job accessibility for that location. The catchment area "floats" from one residential location to another across the study area, and defines the accessibility for all locations. The catchment area may also be defined as a circle (Immergluck, 1998; Wang, 2000) or a fixed travel time range (Wang and Minor, 2002), and the concept remains the same.

Figure 5.1 uses an example to illustrate the method. For simplicity, assume that each demand location (e.g., tract) has only one resident at its centroid and the capacity of each supply location is also 1. Assume that a circle around the centroid of a residential location defines its catchment area. Accessibility in a tract is defined as the supply-to-demand ratio within its catchment area. For instance, within the catchment area of tract 2, total supply is 1 (i.e., only *a*) and total demand is 7. Therefore, accessibility at tract 2 is the supply–demand ratio, i.e., 1/7. The circle floats from one centroid to another while its radius remains the same. The catchment area of tract 11 contains a total supply of 3 (i.e., *a*, *b*, and *c*) and a total demand of 7, and thus the accessibility at tract 11 is 3/7. Note that the ratio is based on the floating catchment area and not confined by the boundary of an administrative unit.

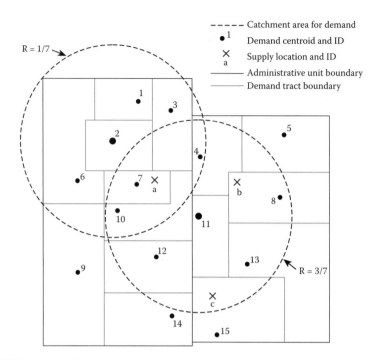

FIGURE 5.1 An earlier version of the FCA method.

The above example can also be used to explain the fallacies of this simple FCA method. It assumes that services within the catchment area are fully available to residents within that catchment area. However, the distance between a supply and a demand within the catchment area may exceed the threshold distance (e.g., in Figure 5.1, the distance between 13 and *a* is greater than the radius of the catchment area of tract 11). Furthermore, the supply at *a* is within the catchment of tract 2, but may not be fully available to serve demands within the catchment, as it is also reachable by tract 11. This points out the need to discount the availability of a supplier by the intensity of competition for its service of surrounding demands.

5.2.2 TWO-STEP FLOATING CATCHMENT AREA (2SFCA) METHOD

A method developed by Radke and Mu (2000) overcomes the above fallacies. It repeats the process of floating catchment twice (once on supply locations and once on demand locations) and is therefore referred to as the *two-step floating catchment area* (2SFCA) *method* (Luo and Wang, 2003).

First, for each supply location *j*, search all demand locations (*k*) that are within a threshold travel distance (d_0) from location *j* (i.e., catchment area *j*) and compute the supply-to-demand ratio R_j within the catchment area:

$$R_j = \frac{S_j}{\sum_{k \in \{d_{kj} \leq d_0\}} D_k}$$

where d_{kj} is the distance between k and j, D_k is the demand at location k that falls within the catchment (i.e., $d_{kj} \leq d_0$), and S_j is the capacity of supply at location j.

Next, for each demand location i, search all supply locations (j) that are within the threshold distance (d_0) from location i (i.e., catchment area i) and sum up the supply-to-demand ratios R_j at those locations to obtain the accessibility A_i^F at demand location i:

$$A_i^F = \sum_{j \in \{d_{ij} \leq d_0\}} R_j = \sum_{j \in \{d_{ij} \leq d_0\}} \left(\frac{S_j}{\sum_{k \in \{d_{kj} \leq d_0\}} D_k} \right) \qquad (5.2)$$

where d_{ij} is the distance between i and j, and R_j is the supply-to-demand ratio at supply location j that falls within the catchment centered at i (i.e., $d_{ij} \leq d_0$). A larger value of A_i^F indicates a better accessibility at a location.

The first step above assigns an initial ratio to each service area centered at a supply location as a measure of supply availability (or crowdedness). The second step sums up the initial ratios in the overlapped service areas to measure accessibility for a demand location, where residents have access to multiple supply locations. The method considers interaction between demands and supplies across areal unit borders and computes an accessibility measure that varies from one location to another. Equation 5.2 is basically the ratio of supply to demand (filtered by a threshold distance or filtering window twice).

Figure 5.2 uses the same example to illustrate the 2SFCA method. Here we use travel time instead of straight-line distance to define catchment area. The catchment area for supply a has one supply and eight residents, and thus carries a supply-to-demand ratio of 1/8. Similarly, the ratio for catchment b is 1/4; and for catchment c, 1/5. The resident at tract 3 has access to a only, and the accessibility at tract 3 is equal to the supply-to-demand ratio at a (the only supply location), i.e., $R_a = 0.125$. Similarly, the resident at tract 5 has access to b only, and thus its accessibility is $R_b = 0.25$. However, the resident at 4 can reach both supplies a and b (shown in an area overlapped by catchment areas a and b), and therefore enjoys a better accessibility (i.e., $R_a + R_b = 0.375$). Note that supply a or b can reach tract 4 within the threshold travel time, and on the other side, tract 4 can reach both supply a and b within the same threshold.

The catchment drawn in the first step is centered at a supply location, and thus the travel time between the supply and any demand within the catchment does not exceed the threshold. The catchment drawn in the second step is centered at a demand location, and all supplies within the catchment contribute to the supply–demand ratios at that demand location. The method overcomes the fallacies in the earlier FCA methods. Equation 5.2 is basically a ratio of supply to demand, with only selected supplies and demands entering the numerator and denominator, and the selections are based on a threshold distance or time within which supplies and demands interact. Travel time should be used if distance is a poor measure of travel impedance (e.g., in areas where roads are unevenly distributed and travel speeds vary to a great extent).

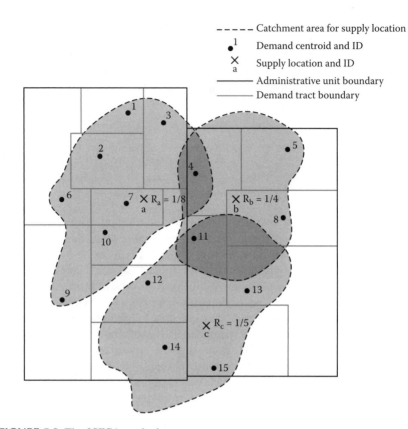

FIGURE 5.2 The 2SFCA method.

The method can be implemented in ArcGIS using a series of join and sum functions. The detailed procedures are explained in Section 5.4.

5.3 THE GRAVITY-BASED METHOD

5.3.1 GRAVITY-BASED ACCESSIBILITY INDEX

The 2SFCA method draws an artificial line (say, 15 miles or 30 minutes) between an accessible and inaccessible location. Supplies within that range are counted equally regardless of the actual travel distance or time (e.g., 2 vs. 12 miles). Similarly, all supplies beyond that threshold are defined as inaccessible, regardless of any differences in travel distance or time. The gravity model rates a nearby supply more accessibly than a remote one, and thus reflects a continuous decay of access in distance.

The potential model in Equation 5.1 considers only the supply side, not the demand side (i.e., competition for available supplies among demands). Weibull (1976) improved the measurement by accounting for competition for services among residents (demands). Joseph and Bantock (1982) applied the method to assess health care accessibility. Shen (1998) and Wang (2001b) used the method for evaluating job accessibility. The gravity-based accessibility measure at location i can be written as

$$A_i^G = \sum_{j=1}^{n} \frac{S_j d_{ij}^{-\beta}}{V_j}, \text{ where } V_j = \sum_{k=1}^{m} D_k d_{kj}^{-\beta} \tag{5.3}$$

A_i^G is the gravity-based index of accessibility, where n and m are the total numbers of supply and demand locations, respectively, and the other variables are the same as in Equations 5.1 and 5.2. Compared to the primitive accessibility measure based on the Hansen model, A_i^H, A_i^G discounts the availability of a physician by the service competition intensity at that location, V_j, measured by its population potential. A larger A_i^G implies better accessibility.

This accessibility index can be interpreted like the one defined by the 2SFCA method. It is essentially the ratio of supply S to demand D, each of which is weighted by travel distances or time to a negative power. The total accessibility scores (i.e., sum of individual accessibility indexes multiplied by corresponding demand amounts), by either the 2SFCA or the gravity-based method, are equal to the total supply. Alternatively, the weighted average of accessibility in all demand locations is equal to the supply-to-demand ratio in the whole study area (see Appendix 5 for a proof).

5.3.2 COMPARISON OF THE 2SFCA AND GRAVITY-BASED METHODS

A careful examination of the two methods further reveals that the two-step floating catchment area (2SFCA) method is merely a special case of the gravity-based method.

The 2SFCA method treats distance (time) impedance as a dichotomous measure; i.e., any distance (time) within a threshold is equally accessible, and any distance (time) beyond the threshold is equally inaccessible. Using d_0 as the threshold travel distance (time), distance or time can be recoded as

1. d_{ij} (or d_{kj}) = ∞ if d_{ij} (or d_{kj}) > d_0
2. d_{ij} (or d_{kj}) = 1 if d_{ij} (or d_{kj}) $\leq d_0$

For any $\beta > 0$ in Equation 5.3, we have

1. $d_{ij}^{-\beta}$ (or $d_{kj}^{-\beta}$) = 0 when d_{ij} (or d_{kj}) = ∞
2. $d_{ij}^{-\beta}$ (or $d_{kj}^{-\beta}$) = 1 when d_{ij} (or d_{kj}) = 1

In case 1, S_j or P_k will be excluded by being multiplied by zero, and in case 2, S_j or P_k will be included by being multiplied by 1. Therefore, Equation 5.3 is regressed to Equation 5.2, and thus the 2SFCA measure is just a special case of the gravity-based measure. Considering that the two methods have been developed in different fields for a variety of applications, this proof validates their rationale for capturing the essence of accessibility measures.

In the 2SFCA method, a larger threshold distance or time reduces variability of accessibility across space, and thus leads to stronger spatial smoothing (Fotheringham et al., 2000, p. 46; also see Chapter 3). In the gravity-based method, a lower value of travel friction coefficient β leads to a lower variance of accessibility scores, and thus stronger spatial smoothing. The effect of a larger threshold travel time in the 2SFCA method is equivalent to that of a smaller travel friction coefficient in the gravity-based method. Indeed, a lower β value implies that travel distances or times matter less and people would travel farther to see a physician.

The gravity-based method seems to be more theoretically sound than the 2SFCA method. However, the 2SFCA method may be a better choice in some cases for two reasons. First, the gravity-based method tends to inflate accessibility scores in poor-access areas, compared to the 2SFCA method, but the poor-access areas are usually the areas of most interest to many public policy makers. Second, the gravity-based method also involves more computation and is less intuitive. In particular, finding the value of the distance friction coefficient β requires actual travel data and is difficult and often infeasible to derive, as such data are often unavailable or costly to obtain.

5.4 CASE STUDY 5: MEASURING SPATIAL ACCESSIBILITY TO PRIMARY CARE PHYSICIANS IN THE CHICAGO REGION

This case study is simplified from a funded project[2] published in Luo and Wang (2003) and other related articles. The funded project utilizes GIS techniques to implement spatial accessibility measures and integrates with aspatial factors to define physician shortage areas in Illinois. The objective is to help the U.S. Department of Health and Human Services improve the practice of designating *health professional shortage areas* (HPSAs), currently a case-by-case manual process in most states.

The study area is identical to that of case study 4A (Section 4.3): 10 Illinois counties in the Chicago consolidated metropolitan statistical area (CMSA). See the inset in Figure 5.4. The following datasets are provided for this project:

1. A shapefile `chitrtcent` for census tract centroids, with the field `popu` representing population extracted from the 2000 census
2. A shapefile `chizipcent` for zip code area centroids, with the field `doc00` for the number of primary care physicians in each zip code area based on the 2000 Physician Master File of the American Medical Association (AMA)

A shapefile `county10` for the 10 counties is also provided for reference. Other additional datasets are needed if the optional tasks are to be accomplished. This project simplifies and skips some steps for emphasizing the implementation of accessibility measures. Two of these steps are provided as options for readers with interests in these tasks. The first optional task is to compute population-weighted centroids of census tracts and zip code areas to represent the locations of population and physicians more accurately (see Part 1, step 1, below). One may simply use geographic centroids instead of population-weighted centroids, but the differences between them may be significant, particularly in rural or peripheral suburban areas, where population or business tend to concentrate in limited space. Implementing this task would require the block-level population census data and corresponding spatial data. Both datasets can be downloaded from the ESRI website as instructed in Section 1.2. For convenience, the calculated population-weighted centroids for census tracts and zip code areas are provided in the shapefiles `chitrtcent` and `chizipcent`, respectively. The second optional task is to estimate travel times (see Section 5.4.1, step 11). The task utilizes the road network TIGER files, which can also be downloaded from the ESRI website. This case study will simply use straight-line distances to illustrate the methods.

The project will only use the two point–based shapefiles for model computation: `chitrtcent` for the demand side (population) and `chizipcent` for the supply side (physicians). The polygon coverage `chitrt` already used in case study 4A will be used here for mapping the results.

5.4.1 PART 1: IMPLEMENTING THE 2SFCA METHOD

1. Optional: *Generating population-weighted centroids of census tracts and zip code areas*: After the block-level data are downloaded and processed, a spatial layer of all blocks in the 10-county region is created, and its attribute table contains the population data. Add the *x-y* coordinates to the attribute table and overlay the block layer with that of tracts (zip code area) to identify which blocks fall within each tract (zip code area). Compute the weighted *x-y* coordinates based on block-level population data such as

$$x_c = (\sum_{i=1}^{n_c} p_i x_i) / (\sum_{i=1}^{n_c} p_i) \text{ and } y_c = (\sum_{i=1}^{n_c} p_i y_i) / (\sum_{i=1}^{n_c} p_i)$$

where x_c and y_c are the *x* and *y* coordinates of the *population-weighted centroid* of a census tract, respectively; x_i and y_i are the *x* and *y* coordinates of the *i*th block centroid within that census tract, respectively; p_i is the population at the *i*th census block within that census tract; and n_c is the total number of blocks within that census tract.

The computation can be implemented in ArcToolbox by utilizing Spatial Statistics Tools > Measuring Geographic Distribution > Mean Center. For generating census tract centroids, in the dialog window, choose the layer of census block centroids as the Input Feature Class, enter `chitrtcent` as the name for Output Feature Class, choose the population field as the Weight Field and the census tract ID (e.g., tract's STFID code) as the Case Field. For generating zip code centroids, one needs to use a map overlay tool (see Section 1.3), generate a layer with census blocks corresponding to zip code areas, and then use the above Mean Center tool. The Output Feature Class is named `chizipcent`, and the Case Field is "zip code."

2. *Computing distances between census tracts and zip code areas*: Based on the layer `chitrtcent` for census tract centroids and the layer `chizipcent` for zip code centroids (either from step 1 or directly from the CD), use the Point Distance tool in ArcToolbox to compute the distance table `DistAll.dbf` for Euclidean distances between population locations (`chitrtcent`) and physician locations (`chizipcent`).

3. *Extracting distances within a threshold*: Based on the distance table `DistAll.dbf`, select the records ≤ 20 miles (i.e., 32,180 m) and export to a table `Dist20mi.dbf`. The new distance table only includes those distances within the threshold of 20 miles,[3] and thus implements the selection conditions $i \in \{d_{ij} \leq d_0\}$ and $k \in \{d_{kj} \leq d_0\}$ in Equation 5.2.

4. *Attaching population and physician data to the distance table*: Join the attribute table of physicians (`chizipcent`) and that of population (`chitrtcent`) to the distance table `Dist20mi.dbf` by corresponding zip code areas and census tracts, respectively.

5. *Summing population around each physician location*: Based on the updated table `Dist20mi.dbf`, generate a new table `DocAvl.dbf` by summing population by physician locations. The field `sum_popu` is the total population within the threshold distance from each physician location, i.e., calibrating the term $\displaystyle\sum_{k \in \{d_{kj} \leq d_0\}} D_k$ in Equation 5.2.

6. *Computing initial physician-to-population ratio at each physician location*: Join the attribute table of physicians (`chizipcent`) to `DocAvl.dbf`, add a field `docpopR`, and compute it as `docpopR = 1000*doc00/ sum_popu`. This assigns an initial physician-to-population ratio to each physician location, indicating the physician availability per 1000 residents. This step implements the term $\displaystyle\frac{S_j}{\sum_{k \in \{d_{kj} \leq d_0\}} D_k}$ in Equation 5.2.

7. *Attaching initial physician-to-population ratios to distance table*: Join the updated table `DocAvl.dbf` to the table `Dist20mi.dbf` by physician locations.

8. *Summing up physician-to-population ratios by population locations*: Based on the updated `Dist20mi.dbf`, sum the initial physician-to-population ratios `docpopR` by population locations to yield a new table `TrtAcc.dbf`. The field `sum_docpopR` in the table `TrtAcc.dbf` sums up availability of physicians that are reachable from each residential location, and thus yields the accessibility A_i^F in Equation 5.2.

   Figure 5.3 illustrates the process of table joins and computation from steps 4 to 8.

9. *Mapping accessibility*: Join the table `TrtAcc.dbf` to the census tract centroid shapefile and then to the census tract polygon coverage for mapping (see note 6 in Chapter 4).

   Figure 5.4 shows the result using the 20-mile threshold. The accessibility exhibits a monocentric pattern, with the highest score near the city center and declining outward. See more discussion at the end of the section.

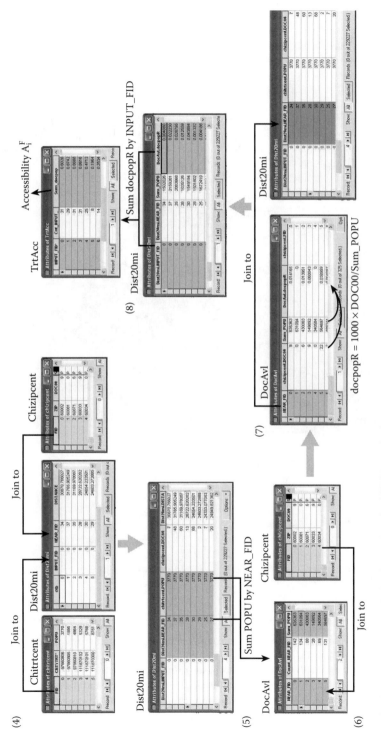

FIGURE 5.3 Procedures in implementing the 2SFCA method.

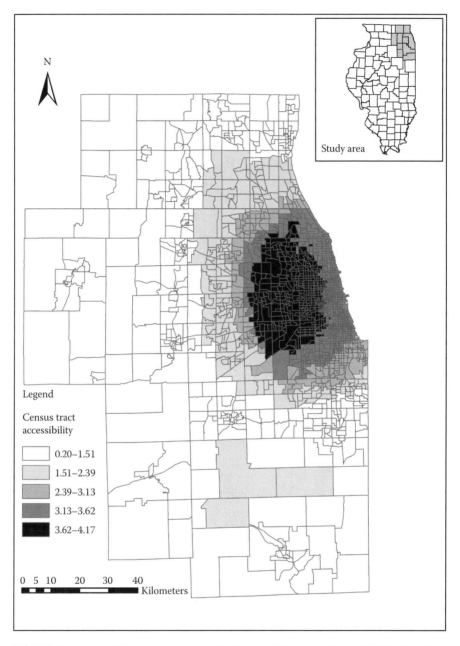

FIGURE 5.4 Accessibility to primary care physician in Chicago region by 2SFCA (20 mile).

10. *Sensitivity analysis using various threshold distances*: A sensitivity analysis can be conducted to examine the impact of threshold distance. For instance, the study can be repeated through steps 3 to 9 using threshold distances of 15, 10, and 5 miles, and results can be compared.

TABLE 5.1
Travel Speed Estimations in the Chicago Region

Category (CFCC)	Area	Speed Limit (mph)
Interstate hwy. (A11–A18)	Urban and suburban	55
	Rural	65
U.S. and state hwy. and some county hwy. (A21–A38)	Urban	35
	Suburban	45
	Rural	55
Local roads (A41–A48 and others)	Urban	20
	Suburban	25
	Rural	35

11. Optional: *Estimating travel times*: Based on the population densities of
census tracts, a layer is created with various categories of density areas.
This layer of density categories is then overlaid with the road network so
that each road segment can be assigned a travel speed according to:
a. Its level based on the CFCC code
b. Density type (i.e., urban, suburban, or rural, as shown in Table 5.1) in
order to account for the congestion effect in different density areas
See Table 5.1 for the recommended rules. Travel speeds are used to define
impedance values in the network shortest-route computation. Follow the
steps in Section 2.3.2 to compute the travel time table between census
tract centroids and zip code area centroids. The remaining steps for computing accessibility are the same as in steps 3 to 9.
Figure 5.5 shows the result by 2SFCA using a threshold of 30-minute
travel time. While the general pattern is consistent with Figure 5.4 based
on straight-line distances, it shows the areas of high accessibility stretching along expressways (particularly high around the intersections).
Different threshold travel times can also be tested for sensitivity analysis.
Table 5.2 shows the result by varying the threshold time from 20 to
50 minutes.

5.4.2 Part 2: Implementing the Gravity-Based Model

The process of implementing the gravity-based model is similar to that of the 2SFCA
method. The following only points out the differences from Part 1.

The gravity model utilizes all distances (times) or the distances (times) up to a
maximum (i.e., an upper limit for one to visit a primary care physician). Therefore,
step 3 in Part 1 for extracting distances within a threshold is skipped, and computation will be based on the original distance table `DistAll.dbf`.

In step 5, add a field, `PPotent`, to the table `DistAll.dbf` and compute it as
`PPotent = popu*distance^(-1)` (assuming a travel friction coefficient $\beta = 1.0$
here). Based on the updated distance table, generate a new table `DocAvlg.dbf`
(adding *g* to the table name to differentiate from those in Part 1) by summing `PPotent`

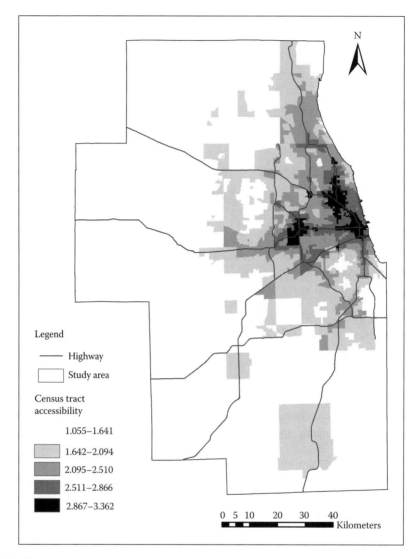

FIGURE 5.5 Accessibility to primary care physician in Chicago region by 2SFCA (30 minute).

by physician locations. The field $\texttt{Sum\_PPotent}$ calibrates the term $V_j = \sum_{k=1}^{m} D_k d_{kj}^{-\beta}$ in Equation 5.3, which is the population potential for each physician location.

In step 6, join the attribute table of physicians ($\texttt{chizipcent}$) to $\texttt{DocAvlg.dbf}$.

In step 7, join the updated table $\texttt{DocAvlg.dbf}$ to the table $\texttt{DistAll.dbf}$ by physician locations, add a field R, and compute it as $\texttt{R = 1000*doc00*}$ $\texttt{distance\^{}(-1)/sum\_PPotent}$. This computes the term $\dfrac{S_j d_{ij}^{-\beta}}{V_j}$ in Equation 5.3.

TABLE 5.2
Comparison of Accessibility Measures

Method	Parameter	Min.	Max.	Std. Dev.	Mean	Weighted Mean
	20 min	0	14.088	2.567	2.721	
	25 min	0	7.304	1.548	2.592	
	30 min	0.017	5.901	1.241	2.522	
2SFCA (threshold time)	35 min	0.110	5.212	1.113	2.498	
	40 min	0.175	4.435	1.036	2.474	
	45 min	0.174	4.145	0.952	2.446	
	50 min	0.130	3.907	0.873	2.416	
						2.647
	$\beta = 0.6$	1.447	2.902	0.328	2.353	
	$\beta = 0.8$	1.236	3.127	0.430	2.373	
	$\beta = 1.0$	1.055	3.362	0.527	2.393	
Gravity-based method	$\beta = 1.2$	0.899	3.606	0.618	2.413	
	$\beta = 1.4$	0.767	3.858	0.705	2.433	
	$\beta = 1.6$	0.656	4.116	0.787	2.452	
	$\beta = 1.8$	0.562	4.380	0.863	2.470	

In step 8, on the updated `DistAll.dbf`, sum up R by population locations and name the output table `TrtAccg.dbf`. The field `sum_R` is the gravity-based accessibility A_i^G in Equation 5.3. The result is shown in Figure 5.6, which is based on estimated travel times between census tracts and zip code areas and uses $\beta = 1.0$.

In step 10, the sensitivity analysis can be conducted by varying the β value, e.g., using β in the range of 0.6 to 1.8 by an increment of 0.2. See the results summarized in Table 5.2.

5.5 DISCUSSION AND REMARKS

As shown in Figure 5.4, Figure 5.5, and Figure 5.6, the highest accessibility is generally found in the central city and declines outward to suburban and rural areas. This is most evident in Figure 5.4, when straight-line distances are used. The main reason is the concentration of major hospitals in the central city. When travel times are used, Figure 5.5 and Figure 5.6 show similar patterns where areas of better accessibility stretch along the highways. In general, the gravity-based method uses all distances and thus has a stronger spatial smoothing effect than the 2SFCA method, as shown in Figure 5.5 and Figure 5.6.

Table 5.2 is compiled for comparison of various accessibility measures. As the threshold time in the 2SFCA method increases from 20 to 50 minutes, the variance (or standard deviation) of accessibility measures declines (also the range from minimum to maximum shrinks), and thus leads to stronger spatial smoothing. As the travel friction coefficient β in the gravity-based method increases from 0.6 to 1.8, the variance of accessibility measures increases, which is equivalent to the effect of smaller thresholds in the 2SFCA method. In general (within the reasonable parameter

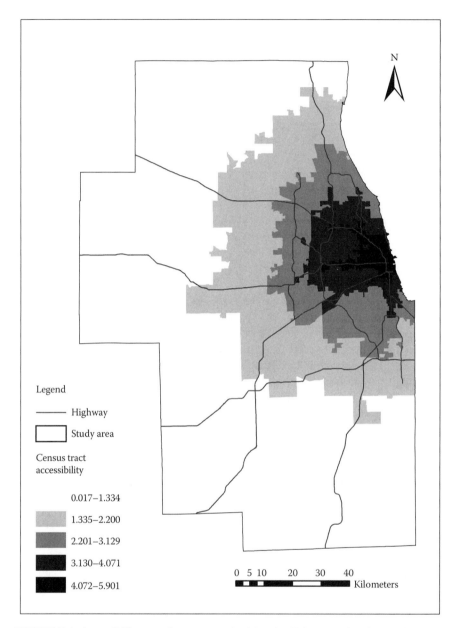

FIGURE 5.6 Accessibility to primary care physician in Chicago region by gravity-based method ($\beta = 1$).

ranges), the gravity-based method has a stronger spatial smoothing effect than the 2SFCA method. This confirms the discussion in Section 5.3. The simple mean of accessibility scores varies slightly by different methods using different parameters. However, the weighted mean remains the same, confirming the property proven in Appendix 5.

The 2SFCA using a larger threshold time generates accessibility with a smaller variance, an effect similar to the gravity-based method using a smaller friction coefficient β. For instance, comparing the accessibility scores by the 2SFCA method with $d_0 = 50$ to the scores by the gravity-based method with $\beta = 1.8$, they have a similar variance (see Table 5.2). However, the distribution of accessibility scores differs. Figure 5.7a shows the distribution by the 2SFCA method (skewed toward high scores). Figure 5.7b shows the distribution by the gravity-based method (a more evenly distributed bell shape). Figure 5.7c plots them in one graph, showing that the gravity-based method tends to inflate the scores in low-accessibility areas.

Like any accessibility study, results near the borders of the study area need to be interpreted with caution because of *edge effects* (Section 3.1). In other words, physicians outside of the study area should also contribute to accessibility of residents near the borders, but are not accounted for in this study. Also note that only spatial accessibility is considered in this study. Residents in areas with high scores of spatial accessibility (e.g., in the inner city) may not actually enjoy good access to health care. Aspatial factors, as discussed in Section 5.1, also play important roles in affecting accessibility. Indeed, the U.S. Department of Health and Human Services designates two types of health professional shortage areas (HPSAs): *geographic areas* and *population groups*. Generally, geographic-area HPSA is intended to capture spatial accessibility and population-group HPSA accounts for aspatial factors. See Wang and Luo (2005) for an approach integrating spatial and aspatial factors in assessing health care access.

Once the distance matrix is generated, one may implement both the accessibility measures in a statistical package like SAS or writing a computing program. However, none is as simple and straightforward as shown in this case study implemented in ArcGIS. The process may be automated in a simple script, which will be convenient for sensitivity analysis.

Accessibility is a common issue in many studies at various scales. For example, a local community may be interested in examining the accessibility to children's playgrounds, identifying underserved areas, and designing a plan of constructing new playgrounds or expanding existing ones. One may also measure the accessibility to public golf courses in a metropolitan area and examine which areas suffer from poor access. The following introduces a case study of measuring job accessibility, an issue with important implications in spatial mismatch, welfare reform, and others.

Data needed for measuring job accessibility include (1) job distribution (supply side), (2) resident worker distribution (demand side), and (3) transportation network (linking the supply and demand). Attribute data for both jobs and resident workers can be extracted from the Census Transportation Planning Package (CTPP). For instance, the 2000 CTPP data are available at http://www.fhwa.dot.gov/ctpp/dataprod.htm. CTPP Part 1 contains residence tables for mapping resident workers, and Part 2 has place-of-work tables for mapping jobs. For microscopic studies (e.g., measuring accessibility within a metropolitan area), the analysis unit is usually traffic analysis zone (TAZ) in 1980 and 1990, and census tract in 2000. Spatial data for TAZs or census tracts can be downloaded from the ESRI data website. Like case study 5, roads extracted from the TIGER line files may be used to approximate the

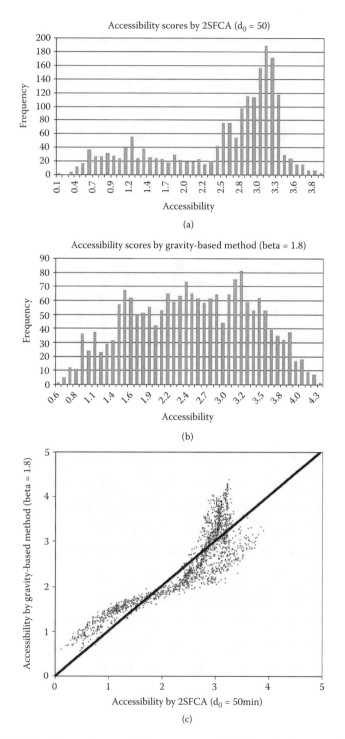

FIGURE 5.7 Comparison of accessibility scores by the 2SFCA and gravity-based methods.

transportation network for defining travel times. Job accessibility can be defined by following techniques and procedures similar to those discussed in Section 5.4.

APPENDIX 5: A PROPERTY FOR ACCESSIBILITY MEASURES

The accessibility index measured in Equation 5.2 or 5.3 has an important property: the total accessibility scores (sum of accessibility multiplied by demand) are equal to the total supply. This implies that the weighted mean of accessibility is equal to the ratio of total supply to total demand in the study area. The following uses measures by the gravity-based method to prove the property (also see Shen, 1998, pp. 363–364). As shown in Section 5.3, the two-step floating catchment area (2SFCA) method in Equation 5.2 is only a special case of the gravity-based method in Equation 5.3, and therefore the proof also applies to the index defined by the 2SFCA method.

Recall the gravity-based accessibility index for demand site i written as

$$A_i^G = \sum_{j=1}^{n} \frac{S_j d_{ij}^{-\beta}}{\sum_{k=1}^{m} D_k d_{kj}^{-\beta}} \tag{A5.1}$$

The total accessibility (TA) is

$$TA = \sum_{i=1}^{m} D_i A_i^G = D_1 A_1^G + D_2 A_2^G + ... + D_m A_m^G \tag{A5.2}$$

Substituting Equation A5.1 into Equation A5.2 and expanding the summation terms yields

$$TA = D_1 \sum_j \frac{S_j d_{1j}^{-\beta}}{\sum_k D_k d_{kj}^{-\beta}} + D_2 \sum_j \frac{S_j d_{2j}^{-\beta}}{\sum_k D_k d_{kj}^{-\beta}} + ... + D_m \sum_j \frac{S_j d_{mj}^{-\beta}}{\sum_k D_k d_{kj}^{-\beta}}$$

$$= \frac{D_1 S_1 d_{11}^{-\beta}}{\sum_k D_k d_{k1}^{-\beta}} + \frac{D_1 S_2 d_{12}^{-\beta}}{\sum_k D_k d_{k2}^{-\beta}} + ... + \frac{D_1 S_n d_{1n}^{-\beta}}{\sum_k D_k d_{kn}^{-\beta}}$$

$$+ \frac{D_2 S_1 d_{21}^{-\beta}}{\sum_k D_k d_{k1}^{-\beta}} + \frac{D_2 S_2 d_{22}^{-\beta}}{\sum_k D_k d_{k2}^{-\beta}} + ... + \frac{D_2 S_n d_{2n}^{-\beta}}{\sum_k D_k d_{kn}^{-\beta}} + ...$$

$$+ \frac{D_m S_1 d_{m1}^{-\beta}}{\sum_k D_k d_{k1}^{-\beta}} + \frac{D_m S_2 d_{m2}^{-\beta}}{\sum_k D_k d_{k2}^{-\beta}} + ... + \frac{D_m S_n d_{mn}^{-\beta}}{\sum_k D_k d_{kn}^{-\beta}}$$

Rearranging the terms, we obtain

$$
TA = \frac{D_1 S_1 d_{11}^{-\beta} + D_2 S_1 d_{21}^{-\beta} + \ldots + D_m S_1 d_{m1}^{-\beta}}{\sum_k D_k d_{k1}^{-\beta}} + \frac{D_1 S_2 d_{12}^{-\beta} + D_2 S_2 d_{22}^{-\beta} + \ldots + D_m S_2 d_{m2}^{-\beta}}{\sum_k D_k d_{k2}^{-\beta}} + \ldots
$$

$$
+ \frac{D_1 S_n d_{1n}^{-\beta} + D_2 S_n d_{2n}^{-\beta} + \ldots + D_m S_n d_{mn}^{-\beta}}{\sum_k D_k d_{kn}^{-\beta}}
$$

$$
= \frac{S_1 \sum_k D_k d_{k1}^{-\beta}}{\sum_k D_k d_{k1}^{-\beta}} + \frac{S_2 \sum_k D_k d_{k2}^{-\beta}}{\sum_k D_k d_{k2}^{-\beta}} + \ldots + \frac{S_n \sum_k D_k d_{kn}^{-\beta}}{\sum_k D_k d_{kn}^{-\beta}}
$$

$$
= S_1 + S_2 + \ldots + S_n
$$

Denoting the total supply in the study area as S (i.e., $S = \sum_{i=1}^{n} S_i$), the above equation shows that $TA = S$; i.e., total accessibility is equal to total supply.

Denoting the total demand in the study area as D (i.e., $D = \sum_{i=1}^{m} D_i$), the weighted average of accessibility is

$$
W = \sum_{i=1}^{m} (\frac{D_i}{D}) A_i^G = (1/D)(D_1 A_1^G + D_2 A_2^G + \ldots + D_m A_m^G) = TA / D = S / D
$$

which is the ratio of total supply to total demand.

NOTES

1. For details and updates, visit http://bphc.hrsa.gov/dsd.
2. The research was supported by the U.S. Department of Health and Human Services, Agency for Healthcare Research and Quality, under Grant 1-R03-HS11764-01.
3. A reasonable threshold distance for a primary care physician is 15 miles (travel distance). We set the search radius at 20 miles (Euclidean distance), roughly equivalent to a travel distance of 30 miles. This is perhaps an upper limit.

6 Function Fittings by Regressions and Application in Analyzing Urban and Regional Density Patterns

Urban and regional studies begin with analyzing the spatial structure, particularly population density patterns. As population serves as both supply (labor) and demand (consumers) in an economic system, the distribution of population represents that of economic activities. Analysis of changing population distribution patterns is a starting point for examining economic development patterns in a city or region. Urban and regional density patterns mirror each other: the *central business district* (CBD) is the center of a city, whereas the whole city itself is the center of a region, and densities decline with distances both from the CBD in a city and from the central city in a region. While the theoretical foundations for declining urban and regional density patterns are different (see Section 6.1), the methods for empirical studies are similar and closely related.

This chapter discusses how we can find a function capturing the density patterns best, and what we can learn about urban and regional growth patterns from this approach. The methodological focus is on function fittings by regressions and related statistical issues. Section 6.1 explains how density functions are used to examine urban and regional structures. Section 6.2 presents various functions for a monocentric structure. Section 6.3 discusses some statistical concerns on monocentric function fittings and introduces nonlinear regression and weighted regression. Section 6.4 examines various assumptions for a polycentric structure and corresponding function forms. Section 6.5 uses a case study in the Chicago region to illustrate the techniques (monocentric vs. polycentric models, linear vs. nonlinear and weighted regressions). The chapter is concluded in Section 6.6 with discussion and a brief summary.

6.1 THE DENSITY FUNCTION APPROACH TO URBAN AND REGIONAL STRUCTURES

6.1.1 STUDIES ON URBAN DENSITY FUNCTIONS

Since the classic study by Clark (1951), there has been great interest in empirical studies of urban population density functions. This cannot be solely explained by the easy availability of data. Many are attracted to the research topic because of its power of revealing urban structure and its solid foundation in economic theory.[1] McDonald (1989, p. 361) considers the population density pattern as "a critical economic and social feature of an urban area."

Among all functions, the *exponential function* or *Clark's model* is the one used most widely:

$$D_r = ae^{br} \qquad\qquad (6.1)$$

where D_r is the density at distance r from the city center (i.e., CBD), a is a constant (the CBD intercept), and b is also a constant for the density gradient. Since the density gradient b is often a negative value, the function is also referred to as the *negative exponential function*. Empirical studies show that it is a good fit for most cities in both developed and developing countries (Mills and Tan, 1980).

The economic model by Mills (1972) and Muth (1969), often referred to as the *Mills–Muth model*, is developed to explain the empirical pattern of urban densities as a negative exponential function. The model assumes a *monocentric structure*: a city has only one center, where all employment is concentrated. Intuitively, as everyone commutes to the city center for work, a household farther away from the CBD spends more on commuting and is compensated by living in a larger-lot house (also cheaper in terms of price per area unit). The resulting population density exhibits a declining pattern with distance from the city center. Appendix 6A shows how the negative exponential urban density function is derived in the economic model. From the deriving process, parameter b in Equation 6.1 is the unit cost of transportation. Therefore, declining transportation costs over time, as a result of improvements in transportation technologies and road networks, lead to a flatter density gradient. This clearly explains that *urban sprawl* and *suburbanization* are mainly attributable to transportation improvements.

However, economic models are "simplification and abstractions that may prove too limiting and confining when it comes to understanding and modifying complex realities" (Casetti, 1993, p. 527). The main criticisms lie in its assumptions of the monocentric city and unit price elasticity for housing, neither of which is supported by empirical studies. Wang and Guldmann (1996) developed a *gravity-based model* to explain the urban density pattern (also see Appendix 6A). The basic assumption of the gravity-based model is that population at a particular location is proportional to its accessibility to all other locations in a city, measured as a gravity potential. Simulated density patterns from the model conform to the negative exponential function when the distance friction coefficient β falls within a certain range ($0.2 \leq \beta \leq 1.0$ in the simulated example). The gravity-based model does not make the restrictive assumptions as in the economic model, and thus implies wide applicability. It also explains two important empirical findings: (1) flattening density gradient over times (corresponding to smaller β) and (2) flatter gradients in larger cities. The economic model explains the first finding well, but not the second (McDonald, 1989, p. 380). Both the economic model and the gravity-based model explain the change of density gradient over time through transportation improvements. Note that both the distance friction coefficient β in the gravity model and the unit cost of transportation in the economic model decline over time.

Earlier empirical studies of urban density patterns are based on the monocentric model, i.e., how population density varies with distance from the city center. It emphasizes the impact of the primary center (CBD) on citywide population distribution. Since the 1970s, more and more researchers recognize the changing urban form from *monocentricity* to *polycentricity* (Ladd and Wheaton, 1991; Berry and Kim, 1993). In addition to the major center in the CBD, most large cities have secondary centers or subcenters, and thus are better characterized as polycentric cities. In a polycentric city, assumptions of whether residents need to access all centers or some of the centers lead to various function forms. Section 6.4 will examine the polycentric models in detail.

6.1.2 STUDIES ON REGIONAL DENSITY FUNCTIONS

The study of regional density patterns is a natural extension to that of urban density patterns as the study area is expanded to include rural areas. The urban population density patterns, particularly the negative exponential function, are empirically observed first, and then explained by theoretical models (either the economic model or the gravity-based model). Even considering the *Alonso's* (1964) *urban land use model* as the precedent of the Mills–Muth urban economic model, the theoretical explanation lags behind the empirical finding on urban density patterns. In contrast, following the rural land use theory by von Thünen (1966, English version), economic models for the regional density pattern by Beckmann (1971) and Webber (1973) were developed before the work of empirical models for regional population density functions by Parr (1985), Parr et al. (1988), and Parr and O'Neill (1989). The city center used in the urban density models remains as the center in regional density models. The declining regional density pattern has a different explanation. In essence, rural residents farther away from a city pay higher transportation costs for the shipment of agricultural products to the urban market and for gaining access to industrial goods and urban services in the city, and are compensated by occupying cheaper, and hence more, land. See Wang and Guldmann (1997) for a recent model.

Similarly, empirical studies of regional density patterns can be based on a monocentric or a polycentric structure. Obviously, as the territory for a region is much larger than a city, it is less likely for physical environments (e.g., topography, weather, and land use suitability) to be uniform across a region than a city. Therefore, population density patterns in a region tend to exhibit less regularity than in a city. An ideal study area for empirical studies of regional density functions would be an area with uniform physical environments, like the "isolated state" in the von Thünen model (Wang, 2001a, p. 233).

Analyzing the function change over time has important implications for both urban and regional structures. For urban areas, we can examine the trend of *urban polarization* vs. *suburbanization*. The former represents an increasing percentage of population in the urban core relative to its suburbia, and the latter refers to a reverse trend, with an increasing portion in the suburbia. For regions, we can identify the process of *centralization* vs. *decentralization*. Similarly, the former

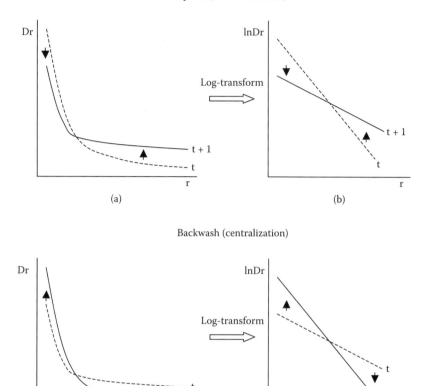

FIGURE 6.1 Regional growth patterns by the density function approach.

refers to the migration trend from peripheral rural to central urban areas, and the latter is the reverse. Both can be synthesized into a framework of core vs. periphery. According to Gaile (1980), economic development in the core (city) impacts the surrounding (suburban and rural) region through a complex set of dynamic spatial processes (i.e., intraregional flows of capital, goods and services, information and technology, and residents). If the processes result in an increase in activity (e.g., population) in the periphery, the impact is *spread*. If the activity in the periphery declines while the core expands, the effect is *backwash*. Such concepts help us understand core–hinterland interdependencies and various relationships between them (Barkley et al., 1996). If the exponential function is a good fit for regional density patterns, the changes can be illustrated as in Figure 6.1, where $t + 1$ represents a more recent time than t. In a monocentric model, we can see the relative importance of the city center; in a polycentric model, we can understand the strengthening or weakening of various centers.

In the reminder of this chapter, the discussion focuses on urban density patterns. However, similar techniques can be applied to studies of regional density patterns.

6.2 FUNCTION FITTINGS FOR MONOCENTRIC MODELS

6.2.1 Four Simple Bivariate Functions

In addition to the exponential function (Equation 6.1) introduced earlier, three other simple bivariate functions for the monocentric structure have often been used:

$$D_r = a + br \tag{6.2}$$

$$D_r = a + b lnr \tag{6.3}$$

$$D_r = ar^b \tag{6.4}$$

Equation 6.2 is a *linear function*, Equation 6.3 is a *logarithmic function*, and Equation 6.4 is a *power function*. Parameter b in all the above four functions is expected to be negative, indicating declining densities with distances from the city center.

Equation 6.2 and 6.3 can be easily estimated by *ordinary least squares* (OLS) *linear regressions*. Equations 6.1 and 6.4 can be transformed to linear functions by taking the logarithms on both sides, such as

$$lnD_r = A + br \tag{6.5}$$

$$lnD_r = A + b lnr \tag{6.6}$$

Equation 6.5 is the log-transform of exponential Equation 6.1, and Equation 6.6 is the log-transform of power Equation 6.4. The intercept A in both Equations 6.5 and 6.6 is just the log-transform of constant a (i.e., $A = lna$) in Equations 6.1 and 6.4. The value of a can be easily recovered by taking the reverse of logarithm, i.e., $a = e^A$. Equations 6.5 and 6.6 can also be estimated by linear OLS regressions. In regressions for Equations 6.3 and 6.6 containing the term lnr, samples should not include observations where $r = 0$ (exactly the city center), to avoid taking logarithms of zero. Similarly, in Equations 6.5 and 6.6 containing the term $lnDr$, samples should not include those where $D_r = 0$ (with zero population).

Take the log-transform of exponential function in Equation 6.5 for an example. The two parameters, intercept A and gradient b, characterize the density pattern in a city. A lower value of A indicates declining densities around the central city; a lower value of b (in terms of absolute value) represents a flatter density pattern. Many cities have experienced lower intercept A and flatter gradient b over time, representing a common trend of urban sprawl and suburbanization. The changing pattern is similar to Figure 6.1a, which also depicts decentralization in the context of regional growth patterns.

6.2.2 OTHER MONOCENTRIC FUNCTIONS

In addition to the four simple bivariate functions discussed above, three other functions are also used widely in the literature. One was proposed by Tanner (1961) and Sherratt (1960) independently, commonly referred to as the *Tanner–Sherratt model*. The model is written as

$$D_r = ae^{br^2} \tag{6.7}$$

where the density D_r declines exponentially with distance squared, r^2.

Newling (1969) incorporated both Clark's model and the Tanner–Sherratt model and suggested the following model:

$$D_r = ae^{b_1r+b_2r^2} \tag{6.8}$$

where the constant term b_1 is most likely to be positive and b_2 negative, and other notations remain the same. In *Newling's model*, a positive b_1 represents a *density crater* around the CBD, where population density is comparatively low due to the presence of commercial and other nonresidential land uses. According to Newling's model, the highest population density does not occur at the city center, but rather at a certain distance away from the city center.

The third model is the *cubic spline function* used by some researchers (e.g., Anderson, 1985; Zheng, 1991) in order to capture the complexity of urban density pattern. The function is written as

$$D_x = a_1 + b_1(x - x_0) + c_1(x - x_0)^2 + d_1(x - x_0)^3 + \sum_{i=1}^{k} d_{i+1}(x - x_i)^3 Z_i^* \tag{6.9}$$

where x is the distance from the city center, D_x is the density there, x_0 is the distance of the first density peak from the city center, x_i is the distance of the ith knot from the city center (defined by either the second, third, etc., density peak or simply even intervals across the whole region), and Z_i^* is a dummy variable (= 0, if x is inside the knot; = 1, if x is outside of the knot).

The cubic spline function intends to capture more fluctuations of the density pattern (e.g., local peaks in suburban areas), and thus cannot be strictly defined as a monocentric model. However, it is still limited to examining density variations related to distance from the city center regardless of directions, and thus assumes a *concentric* density pattern.

6.2.3 GIS AND REGRESSION IMPLEMENTATIONS

The density function analysis only uses two variables: one is Euclidean distance r from the city center, and the other is the corresponding population density D_r.

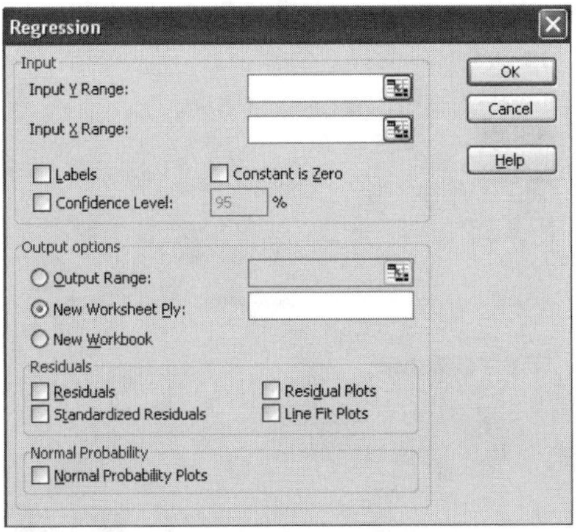

FIGURE 6.2 Excel dialog window for regression.

Euclidean distances from the city center can be obtained using the techniques explained in Section 2.1. Identifying the city center requires knowledge of the study area and is often defined as a commonly recognized landmark site by the public. In the absence of a commonly recognized city center, one may use the local government center[2] or the location with the highest level of job concentration to define it, or follow Alperovich (1982) to identify it as the point producing the highest R^2 in density function fittings. Density is simply computed as population divided by area size. Area size is a default item in any ArcGIS polygon coverage and can be added in a shapefile (see step 3 in Section 1.2). Once the two variables are obtained in GIS, the dataset can be exported to an external file for regression analysis.

Linear OLS regressions are available in many software packages. For example, one may use the widely available Microsoft Excel. Make sure that the Analysis ToolPak is installed in Excel. Open the distance and density data as an Excel workbook, add two new columns to the workbook (e.g., `lnr` and `lnDr`), and compute them as the logarithms of distance and density, respectively. Select Tools from the main menu bar > Data Analysis > Regression to activate the regression dialog window shown in Figure 6.2. By defining the appropriate data ranges for X and Y variables, Equations 6.2, 6.3, 6.5, and 6.6 can be all fitted by OLS linear regressions in Excel. Note that Equations 6.5 and 6.6 are the log-transformations of exponential Equation 6.1 and power Equation 6.4, respectively. Based on the results, Equation 6.1 or 6.4 can be easily recovered by computing the coefficient $a = e^A$ and the coefficient b unchanged.

Alternatively, one may use the Chart Wizard in Excel to obtain the regression results for all four bivariate functions directly. First, use the Chart Wizard to draw a graph depicting how density varies with distance. Then click the graph and choose Chart from the main menu > Add Trendline to activate the dialog window shown in Figure 6.3. Under the menu Type, all four functions (linear, logarithmic, exponential, and power) are available for selection. Under the menu Options, choose "Display equation on chart" and "Display R-squared value on chart" to have regression results shown on the graph. The Add Trendline tool outputs the regression

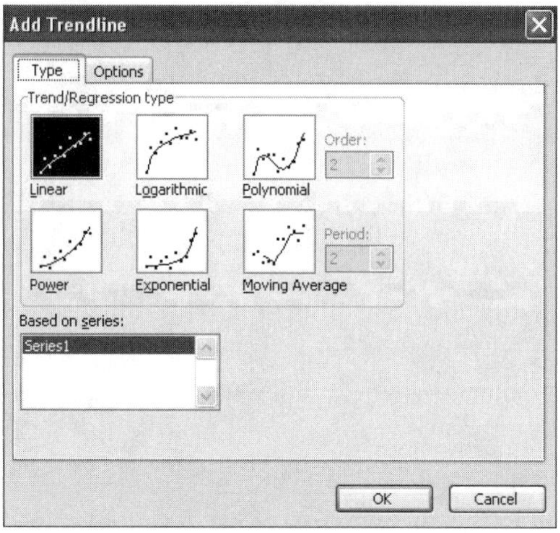

FIGURE 6.3 Excel dialog window for adding trend lines.

TABLE 6.1
Linear Regressions for a Monocentric City

Models	Function Used in Regression	Original Function	X Variable(s)	Y Variable	Restrictions
Linear	$D_r = a + br$	Same	r	D_r	None
Logarithmic	$D_r = a + b\ln r$	Same	$\ln r$	D_r	$r \neq 0$
Power	$\ln D_r = A + b\ln r$	$D_r = ar^b$	$\ln r$	$\ln D_r$	$r \neq 0$ and $D_r \neq 0$
Exponential	$\ln D_r = A + br$	$D_r = ae^{br}$	r	$\ln D_r$	$D_r \neq 0$
Tanner–Sherratt	$\ln D_r = A + br^2$	$D_r = ae^{br^2}$	r^2	$\ln D_r$	$D_r \neq 0$
Newling's	$\ln D_r = A + b_1 r + b_2 r^2$	$D_r = ae^{b_1 r + b_2 r^2}$	r, r^2	$\ln D_r$	$D_r \neq 0$

results for the four original bivariate functions without log-transformations, but does not report as many statistics as the Regression tool. The regression results reported here are based on *linear* OLS regressions by using the log-transform Equations 6.5 and 6.6 (though the computation is done internally). This is different from nonlinear regressions, which will be discussed in the next section.

Both the Tanner–Sherratt model (Equation 6.7) and Newling's model (Equation 6.8) can be estimated by linear OLS regression on their log-transformed forms. See Table 6.1. In the Tanner–Sherratt model, the X variable is distance squared (r^2), and in Newling's model, there are two X variables (r and r^2). Newling's model has one more explanatory variable (r^2) than Clark's model (exponential function), and thus always generates a higher R^2 regardless of the significance of the term r^2. In this sense, these two models are not comparable in terms of fitting power. Table 6.1 summarizes the models.

Fitting the cubic spline function (Equation 6.9) is similar to that of other monocentric functions, with some extra work in preparing the data. First, sort the data by the variable distance in an ascending order. Second, define the constant x_0 and calculate the terms $(x - x_0)$, $(x - x_0)^2$, and $(x - x_0)^3$. Third, define the constants x_i (i.e., $x_1, x_2, ...$) and compute the terms $(x - x_i)^3 Z_i^*$. Take one term, $(x - x_1)^3 Z_1^*$, as an example: (1) set the values $= 0$ for those records with $x \leq x_1$, and (2) compute the values $= (x - x_1)^3$ for those records with $x > x_1$. Finally, run a multivariate regression, where the Y variable is density D_x and the X variables are $(x - x_0)$, $(x - x_0)^2$, $(x - x_0)^3$, $(x - x_1)^3 Z_1^*$, $(x - x_2)^3 Z_2^*$, and so on. The cubic spline function contains multiple X variables, and thus its regression R^2 tends to be higher than other models.

6.3 NONLINEAR AND WEIGHTED REGRESSIONS IN FUNCTION FITTINGS

In function fittings for the monocentric structure, two statistical issues deserve more discussion. One is the choice between nonlinear regressions directly on the exponential and power functions vs. linear regressions on their log-transformations (as discussed in Section 6.2). Generally they yield different results since the two have different dependent variables (D_r in nonlinear regressions vs. lnD_r in linear regressions) and imply different assumptions of error terms (Greene and Barnbrock, 1978).

We use the exponential function (Equation 6.1) and its log-transformation (Equation 6.5) to explain the differences. The linear regression on Equation 6.5 assumes multiplicative errors and weights equal *percentage errors* equally, such as

$$D_r = ae^{br + \varepsilon} \tag{6.10}$$

The nonlinear regression on the original function (Equation 6.1) assumes that additive errors and weights all equal *absolute errors* equally, such as

$$D_r = ae^{br} + \varepsilon \tag{6.11}$$

The *ordinary least squares* (OLS) *linear regression* seeks the optimal values of a and b so that residual sum of squares (RSS) is minimized. See Appendix 6B on how the parameters in a bivariate linear function are estimated by the OLS regression. Nonlinear least squares regression has the same objective of minimizing the RSS. For the model in Equation 6.11, it is to minimize

$$RSS = \sum_i (D_i - ae^{br_i})^2$$

where i indexes individual observations. There are several ways to estimate the parameters in nonlinear regression (Griffith and Amrhein, 1997, p. 265), and all methods use iterations to gradually improve guesses. For example, the *modified Gauss–Newton method* uses linear approximations to estimate how RSS changes with small shifts from the present set of parameter estimates. Good initial estimates (i.e., those close to the correct parameter values) are critical in finding a successful non-linear fit. The initialization of parameters is often guided by experience and knowledge of similar studies.

Which is a better method for estimating density functions, linear or nonlinear regression? The answer depends on the emphasis and objective of a study. The linear regression is based on the log-transformation. By weighting equal *percentage errors* equally, the errors generated by high-density observations are scaled down (in terms of percentage). However, the differences between the estimated and observed values in those high-density areas tend to be much greater than those in low-density areas (in terms of absolute value). As a result, the total estimated population in the city can be off by a large margin. On the contrary, the nonlinear regression is to minimize the residual sum of squares directly based on densities instead of their logarithms. By weighting all equal *absolute errors* equally, the regression limits the errors (in terms of absolute value) contributed by high-density samples. As a result, the total estimated population in the city is often closer to the actual value than the one based on linear regression, but the estimated densities in low-density areas may be off by high percentages.

Another issue in estimating urban density functions concerns *randomness of sample* (Frankena, 1978). A common problem for census data (not only in the U.S.) is that high-density observations are many and are clustered in a small area near the city center, whereas low-density ones are fewer and spread in remote areas. In other words, high-density samples may be overrepresented, as they are concentrated within a short distance from the city center, and low-density samples may be underrepre-sented, as they spread across a wide distance range from the city center. A plot of density vs. distance will show many observations in short distances and fewer in long distances. This is referred to as *nonrandomness of sample* and causes biased (usually upward) estimators. A weighted regression can be used to mitigate the problem. Frankena (1978) suggests weighting observations in proportion to their areas. In the regression, the objective is to minimize the weighted residual sum of squares (RSS). Note that R^2 in a weighted regression can no longer be interpreted as a measure of goodness of fit and is called *pseudo-R^2*. See Wang and Zhou (1999) for an example. Some researchers favor samples with uniform area sizes. In case study 6 (Section 6.5.3 in particular), we will also analyze population density functions based on survey townships of approximately same area sizes.

Estimating the nonlinear regression or weighted regression requires the use of advanced statistical software. For example, in SAS, if the DATA step uses DEN to represent density, DIST to represent distance, and AREA to represent area size, the following SAS statements implement the nonlinear regression for the exponential Equation 6.1:

```
proc MODEL; /* procedure for nonlinear regression */

PARMS a 1000 b -0.1; /*initialize parameters */

DEN = a * exp(b * DIST); /* code the fitting function */

fit DEN; /* define the dependent variable */
```

The statement PARMS assigns initial values for parameters a and b in the iteration process. If the model does not converge, experiment with different initial values until a solution is reached.

The weighted regression is run by adding the following statement to the above program:

```
weight AREA; /* define the weight variable */
```

SAS also has a procedure REG to run OLS linear regressions. See the sample SAS program monocent.sas included in the CD for details.

6.4 FUNCTION FITTINGS FOR POLYCENTRIC MODELS

Monocentric density functions simply assume that densities are uniform at the same distance from the city center regardless of directions. Urban density patterns in some cities may be better captured by a polycentric structure. In a polycentric city, residents and businesses value access to multiple centers, and therefore population densities are functions of distances to these centers (Small and Song, 1994, p. 294). Centers other than the primary or major center at the CBD are called subcenters.

6.4.1 POLYCENTRIC ASSUMPTIONS AND CORRESPONDING FUNCTIONS

A polycentric density function can be established under several alternative assumptions:

1. If the influences from different centers are perfectly substitutable so that only the nearest center matters, the city is composed of multiple mono-centric subregions. Each subregion is the proximal area for a center (see Section 4.1), within which various monocentric density functions can be estimated. Taking the exponential function as an example, the model for the subregion around the ith center (CBD or subcenter) is

$$D = a_i e^{b_i r_i} \tag{6.12}$$

where D is the density of an area, r_i is the distance between the area and its nearest center, i, and a_i and b_i $(i = 1, 2, ...)$ are parameters to be estimated.

2. If the influences are complementary so that some access to all centers is necessary, then the polycentric density is the product of those monocentric functions (McDonald and Prather, 1994). For example, the log-transformed polycentric exponential function is written as

$$\ln D = a + \sum_{i=1}^{n} b_i r_i \tag{6.13}$$

where D is the density of an area, n is the number of centers, r_i is the distance between the area and center i, and a and b_i $(i = 1, 2, ...)$ are parameters to be estimated.

3. Most researchers (Griffith, 1981; Small and Song, 1994) believe that the relationship among the influences of various centers is between assumptions 1 and 2, and the polycentric density is the sum of center-specific functions. For example, a polycentric model based on the exponential function is expressed as

$$D = \sum_{i=1}^{n} a_i e^{b_i r_i} \tag{6.14}$$

The above three assumptions are based on Heikkila et al. (1989).

4. According to the central place theory, the major center at the CBD and the subcenters play different roles. All residents in a city need access to the major center for higher-order services; for other lower-order services, residents only need to use the nearest subcenter (Wang, 2000). In other words, everyone values access to the major center and access to the nearest center (either the CBD or a subcenter). Using the exponential function as an example, the corresponding model is

$$\ln D = a + b_1 r_1 + b_2 r_2 \tag{6.15}$$

where r_1 is the distance from the major center, r_2 is the distance from the nearest center, and a, b_1, and b_2 are parameters to be estimated.

Figure 6.4 illustrates the different assumptions for a polycentric city. Residents need access to all centers under assumption 2 or 3, but effects are multiplicative in 2 and additive in 3. Table 6.2 summarizes the above discussion.

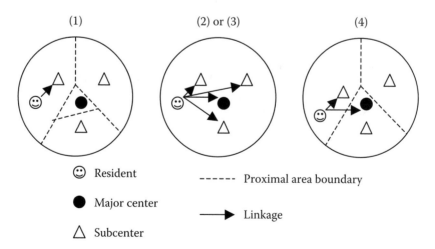

FIGURE 6.4 Illustrations of polycentric assumptions.

TABLE 6.2
Polycentric Assumptions and Corresponding Functions

Label	Assumption	Model (Exponential as an Example)	X Variables	Sample	Estimation Method
1	Only access to the nearest center is needed	$\ln D = A_i + b_i r_i$	Distances r_i from the nearest center i (1 variable)	Areas in a subregion i	Linear regression[a]
2	Access to all centers is necessary (multiplicative effects)	$\ln D = a + \sum_{i=1}^{n} b_i r_i$	Distances from each center (n variables r_i)	All areas	Linear regression
3	Access to all centers is necessary (additive effects)	$D = \sum_{i=1}^{n} a_i e^{b_i r_i}$	Distances from each center (n variables r_i)	All areas	Nonlinear regression
4	Access to CBD and the nearest center is needed	$\ln D = a + b_1 r_1 + b_2 r_2$	Distances from the major and nearest center (2 variables)	All areas	Linear regression[b]

[a] This assumption may be also estimated by nonlinear regression on $D = a_i e^{b_i r_i}$.
[b] This assumption may be also estimated by nonlinear regression on $D = a_1 e^{b_1 r_1} + a_2 e^{b_2 r_2}$.

6.4.2 GIS AND REGRESSION IMPLEMENTATIONS

Analysis of polycentric models requires the identification of multiple centers first. Ideally, these centers should be based on the distribution of employment (e.g., Gordon et al., 1986; Giuliano and Small, 1991; Forstall and Greene, 1998). In addition to traditional choropleth maps, Wang (2000) used surface modeling techniques to generate isolines (contours) of employment density[3] and identified employment centers based on both the contour value (measuring the threshold employment density) and the size of area enclosed by the contour line (measuring the base value of total employment). With the absence of employment distribution data, one may use surface modeling of population density to guide the selection of centers.[4] See Chapter 3 for various surface modeling techniques. Surface modeling is descriptive in nature. Only rigorous statistical analysis of density functions can answer whether the potential centers identified from surface modeling indeed exert influence on surrounding areas and how the influences interact with each other.

Once the centers are identified, GIS prepares the data of distances and densities for analysis of polycentric models. For assumption 1, only the distances from the nearest centers (including the major center) need to be computed by using the Near tool in ArcGIS. For assumption 2 or 3, the distance between each area and every center needs to be obtained by the Point Distance tool in ArcGIS. For assumption 4, two distances are required: the distance between each area and the major center and the distance between each area and its nearest center. The two distances are obtained by using the Near tool in ArcGIS twice. See Section 6.5.2 for details.

Based on assumption 1, the polycentric model is degraded to monocentric functions (Equation 6.12) within each center's proximal area, which can be estimated by the techniques explained in Sections 6.2 and 6.3. Equation 6.13 for assumption 2 and Equation 6.15 for assumption 4 can also be estimated by simple multivariate linear regressions. However, Equation 6.14, based on assumption 3, needs to be estimated by a nonlinear regression, as shown below.

Assuming a model of two centers with `DIST1` and `DIST2` representing the distances from the two centers, respectively, a sample SAS program for estimating Equation 6.14 is similar to the program for estimating Equation 6.4, such as

```
proc model;

parms a1 1000 b1 -0.1 a2 1000 b2 -0.1;

DEN = a1*exp(b1*DIST1)+ a2*exp(b2*DIST2);

fit DEN;
```

6.5 CASE STUDY 6: ANALYZING URBAN DENSITY PATTERNS IN THE CHICAGO REGION

Chicago has been an important study site for urban studies. The classic urban concentric model by Burgess (1925) was based on Chicago and led to a series of studies on urban structure, forming the so-called Chicago School. This case study uses the recent 2000 census data to examine the urban density patterns in the Chicago region. The study area is limited to the core six-county area in Chicago CMSA: Cook (031), DuPage (043), Kane (089), Lake (097), McHenry (111), and Will (197) (county codes in parentheses). The area is smaller than the 10-county study area used in case studies 4A and 5, as we are interested in mostly urbanized areas. See the inset in Figure 6.5. In order to examine the possible *modifiable areal unit problem* (MAUP), the project analyzes the density patterns at both the census tract and survey township levels. The MAUP refers to sensitivity of results to the analysis units for which data are collected or measured, and is well known to geographers and spatial analysts (Openshaw, 1984; Fotheringham and Wong, 1991).

This study also uses the polygon coverage `chitrt` for census tracts in the Chicago 10-county region, which contains the population data (i.e., item `popu`) of 2000. In addition, the following datasets are provided:

1. A shapefile `polycent15` contains 15 centers identified as employment concentrations[5] from a previous study (Wang, 2000).
2. A shapefile `county6` defines the study area of six counties.
3. A shapefile `twnshp` contains 115 survey townships in the study area, providing an alternative areal unit that is relatively uniform in area size.

This study uses ArcGIS for spatial analysis tasks such as distance computation and areal interpolation, Microsoft Excel for simple linear regression and graphs, and SAS for more advanced nonlinear and weighted regressions.

6.5.1 PART 1: FUNCTION FITTINGS FOR MONOCENTRIC MODELS (CENSUS TRACTS)

1. *Data preparation in ArcGIS: extracting study area and CBD location*: Use the tool Selection by Location to select features from `chitrt` that have their centers in `county6`, and export the selected features to a shapefile `cnty6trt`, which contains census tracts in the six-county area. Add a field `popden` to the attribute table of `cnty6trt` and calculate it as popden = 1000000*popu/area.
   Create a shapefile `cnty6trtpt` for tract centroids in the study area from `cnty6trt`.
   Select the point with `cent15_` = 15 (or FID = 14) from the shapefile `polycent15` and export it to a shapefile `monocent`, which identifies the location of CBD (or the only center based on the monocentric assumption).
2. *Mapping population density surface*: Use the surface modeling techniques learned in case study 3B (see Section 3.4) to map the population density surface in the study area. A sample map is shown in Figure 6.5. Note that job centers are not necessarily the peak points of population density, though suburban job centers in general are found to be near the local density peaks. This confirms the necessity of using job distribution data instead of population data to identify urban centers.

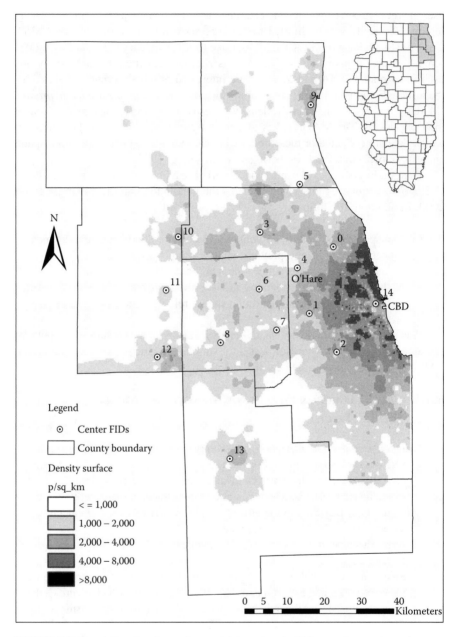

FIGURE 6.5 Population density surface and job centers in Chicago, six-county region.

3. *Computing distances between tract centroids and CBD in ArcGIS*: Use the analysis tool Near to compute distances between tract centroids (cnty6trtpt.shp) and CBD (monocent.shp). In the updated attribute table for cnty6trtpt, the field NEAR_FID identifies the

nearest center (in this case the only point in `monocent`) and the field `NEAR_DIST` is the distance from the center to each tract centroid. Add a field `DIST_CBD` and calculate it as `DIST_CBD = NEAR_DIST/1000`, which is the distance from the CBD in kilometers.[6]

4. *Using the regression tool in Excel to run simple linear regressions*: Make sure that the Analysis ToolPak is installed in Excel. Open `cnty6trtpt.dbf` in Excel and save it as `monodist_trt.xls` in Excel workbook format. Select Tools from the main menu bar > Data Analysis > Regression to activate the dialog window shown in Figure 6.2. Input *Y* range (data range for the variable `popden`) and *X* range (data range for the variable `DIST_CBD`). The linear regression results may be saved in a separate worksheet by clicking the option New Worksheet Ply under Output options. In addition to regression coefficients and R^2, the output includes corresponding standard errors, *t* statistics, and *p* values for the intercept and independent variable.

5. *Using the regression tool in Excel for additional function fittings*: In the file `monodist_trt.xls`, add and compute new fields `dist_sq`, `lndist`, and `lnpopden`, which are the distance squared, logarithm of distance, and logarithm of population density,[7] respectively. For density functions with logarithmic terms (*lnr* or *lnD_r*), the distance (*r*) or density (*D_r*) value cannot be 0. In our case, five tracts have $D_r = 0$. Following common practice, we add 1 to the original variable `popden` when computing *lnD_r*; i.e., the column `lnpopden` is computed as `ln(popden+1)` to avoid taking the logarithm of zero.[8]

   Repeat step 4 to fit the logarithmic, power, exponential, Tanner–Sherratt, and Newling's functions. Use Table 6.1 as a guideline on what data to use for defining the *X* variable(s) and *Y* variable. All regressions, except for Newling's, are simple bivariate models.[9]

   Regression results are summarized in Table 6.3.

6. *Using the trend line tool in Excel for generating graphs and function fittings*: In Excel, use the Chart Wizard to generate a graph (e.g., an *XY* scatter graph) depicting how density varies with distance. Click the graph > choose Chart from the main menu bar > Add Trendline (see Figure 6.3 for a sample dialog window). Under the tab Type, all four functions (linear, logarithmic, exponential, and power) are available for selection. Under the menu Options, choose "Display equation on chart" and "Display R-squared value on chart" to have regression results added to the trend lines. Figure 6.6 shows the exponential trend line superimposed on the *XY* scatter graph of density vs. distance.

   Note that the exponential and power regressions are based on the logarithmic transformation of the original functions, and recovered to the exponential and power function forms by computing the coefficient $a = e^A$ (e.g., for the exponential function, $9157.5 = e^{9.1223}$, see Figure 6.6). Results from the trend line tool are the same as those obtained by the regression tool.

TABLE 6.3
Regressions Based on Monocentric Functions (1837 Census Tracts)

Regression Techniques	Functions	a (or A)	b	R-Squared
Linear	$D_r = a + br$	7187.46	−120.13	0.3237
	$D_r = a + b\ln r$	12071	−2740.91	0.3469
	$\ln D_r = A + b\ln r$	10.09	−0.8135	0.2998
	$\ln D_r = A + br$	8.80	−0.0417	0.3833
	$\ln D_r = A + br^2$	8.29	−0.0005	0.3455
	$\ln D_r = A + b_1 r + b_2 r^2$	8.83	$b_1 = -0.045,\ b_2 = 0.00005$[a]	0.3835
Nonlinear	$D_r = ar^b$	12190.38	−0.3823	0.2455
	$D_r = ae^{br}$	10013.36	−0.0471	0.3912
	$D_r = ae^{br^2}$	8161.09	−0.0019	0.4021
	$D_r = ae^{b_1 r + b_2 r^2}$	6337.17	$b_1 = 0.0577,\ b_2 = -0.0044$	0.4066
Weighted	$D_r = ae^{br}$	9157.49	−0.0603	0.5207[b]

[a] Not significant (all others significant at 0.001).
[b] Pseudo-R-squared is not a measure of goodness of fit.

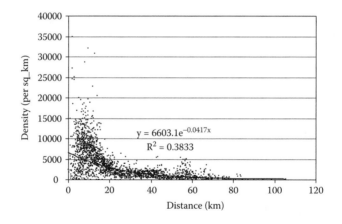

FIGURE 6.6 Density vs. distance exponential trend line (census tracts).

7. *Nonlinear and weighted regressions in SAS*: Nonlinear and weighted regression models need to be estimated in SAS. A sample SAS program monocent.sas is shown in Appendix 6C (also provided in the CD). Prior to running the SAS program, use Excel to extract three columns (DIST_CBD, popden, and area) from monodist_trt.xls and save this as a comma-separate file monodist.csv without the variable names. The SAS program implements all regressions based on monocentric functions. The power, exponential, Tanner–Sherratt, and Newling's functions can all be fit by nonlinear regression. Regression results are summarized in Table 6.3. Among comparable functions, the ones with the

highest R^2 are highlighted. A weighted regression on the exponential function is presented as an example.

6.5.2 Part 2: Function Fittings for Polycentric Models (Census Tracts)

1. *Computing distances between tract centroids and their nearest centers in ArcGIS*: Use the analysis tool Near to compute distances between tract centroids (`cnty6trtpt`) and their nearest center (`polycent15`). In the updated attribute table for `cnty6trtpt`, the fields NEAR_FID and NEAR_DIST are the nearest center and the distance from the center to each tract centroid, respectively.[10] Add a field D_NEARC and calculate it as D_NEARC=NEAR_DIST/1000. Use Excel to extract four columns (POPDEN, DIST_CBD, NEAR_FID, and D_NEARC) from `cnty6trtpt.dbf` and save it as a comma-separate file `dist2near.csv` without the variable names. The file will be used to test the polycentric assumptions 1 and 4.

2. *Computing distances between tract centroids and each center in ArcGIS*: Use the analysis tool Point Distance to compute distances between tract centroids (`cnty6trtpt`) and each of the 15 centers (`polycent15`), and name the output table `polydist.dbf`. Join the attribute table of `cnty6trtpt` to `polydist.dbf` to attach the tract density information, and export it to an external file `tmp.dbf`. In Excel, open `tmp.dbf` and extract attributes popden, INPUT_FID, NEAR_FID, and distance to a comma-delimited text file `dist2cent15.csv` without the variable names. The file will be used to test polycentric assumptions 2 and 3.

3. *Fitting polycentric functions in SAS*: While all linear regressions may be implemented in Excel as shown in Part 1, the program `polycent.sas` provided in the CD fits all functions listed in Table 6.2. Linear regressions are adopted for fitting functions based on assumptions 1, 2, and 4, and nonlinear regression is used for fitting the function based on assumption 3. Assumption 1 implies several simple monocentric functions, each of which is based on the center's proximal area. Assumption 2 leads to a multivariate linear function for the whole study area. See Table 6.4 for the regression results.

   The function based on assumption 3 is a complex nonlinear function with 30 parameters to estimate (two for each center). After numerous trials, the model does not converge and no regression result may be obtained. For illustration, a model with only two centers (center 14 at the CBD and center 4 at O'Hare Airport) is obtained, such as

$$D = 9967.63e^{-0.0456r_{14}} - 6744.84e^{-0.3396r_4} \text{ with } R^2 = 0.3946$$

$$(38.64)(-21.14) \quad (-1.55)(-2.37)$$

TABLE 6.4

Regressions Based on Polycentric Assumptions 1 and 2 (1837 Census Tracts)

Center index i	No. of samples	1: $\ln D = A_i + b_i r_i$ for Center i's Proximal Area			2: $\ln D = a + \sum_{i=1}^{n} b_i r_i$ for the Whole Study Area	
		A_i	b_i	R^2	b_i	
0	184	7.2850***	0.1609***	0.193	−0.1108***	
1	106	7.3964***	0.1529***	0.268	−0.0615*	
2	401	8.3702***	−0.0464***	0.110	−0.0146	
3	76	7.7196***	−0.0515**	0.114	−0.0689**	
4	52	5.6173***	0.3460***	0.260	0.1386***	
5	71	7.1939***	−0.0535**	0.102	0.1155***	$a = 10.98$***
6	51	7.2516***	−0.0010	0.000	−0.0487	
7	46	7.5583***	−0.0694*	0.132	0.1027***	$R^2 = 0.429$
8	58	7.0065***	0.0100	0.003	−0.0341	
9	100	7.4816***	−0.0657***	0.350	−0.0543***	Sample size = 1837
10	86	7.2826***	−0.0576***	0.292	−0.0363*	
11	22	6.6063***	−0.0275	0.012	0.0780**	
12	28	8.0979***	−0.2393***	0.571	−0.0457**	
13	67	7.1424***	−0.0768***	0.412	−0.0044	
14	489	8.1807***	0.0513**	0.015	−0.02762***	

Note: ***, significant at 0.001; **, significant at 0.01; *, significant at 0.05.

The corresponding t values in parentheses indicate that the CBD is far more significant than the O'Hare Airport center.

The function based on assumption 4 is obtained by a linear regression, such as

$$lnD = 8.8584 - 0.0396r_{CBD} - 0.0126r_{cent} \text{ with } R^2 = 0.3946$$

$$(199.36) \quad (-28.54) \quad \quad (-3.28)$$

The corresponding t values in parentheses imply that both distances from the CBD and the nearest center are statistically significant, but the distance from the CBD is far more important.

6.5.3 Part 3: Function Fittings for Monocentric Models (Townships)

1. *Estimating population in townships by areal weighting interpolator in ArcGIS*: Refer to Section 3.6, Part 2, for detailed instructions on areal

weighting interpolator. In ArcToolbox, use the analysis tool Intersect to overlay `cnty6trt` and `twnshp`, and name the output layer `twntrt`. In the attribute table of `twntrt`, the field `area` is the area size for tracts from `cnty6trt`, and the field `area_1` is the area size for townships from `twnshp`. Add a field `area_2` to the attribute table of `twntrt` and update it as the area size for the newly created shapefile `twntrt` (see Section 1.2, Step 3). Add another field `pop_est` and calculate it as `pop_est = popu * area_2/area`. Use the tool Summarize to compute the sum of `pop_est` by the field `rngtwn` (township IDs), and name the output table `twn_pop.dbf`.

2. *Computing densities and distances from CBD for survey townships in ArcGIS*: Join the table `twn_pop.dbf` to the attribute table of `twnshp`, add a field `popden` to the table, and calculate it as `popden = 1000000*sum_pop_est/area`.

   Use the Near tool to obtain the distances of survey townships (`twnshp.shp`) from the CBD (`monocent.shp`). The attribute table of shapefile `twnshp` now contains the variables (`popden` and `NEAR_DIST`) needed for density function fittings.

3. *Monocentric function fittings*: Excel can be used to run linear regression results, similar to Section 6.5.1 steps 4 to 6, but SAS is needed to run nonlinear regressions. For convenience, one may extract needed variables from the attribute table of `twnshp` and feed into a slightly modified SAS program `monocent.sas` (by simply revising the data input statements) to run both the linear and nonlinear regressions. Results are shown in Table 6.5. Given the small sample size, polycentric functions are not tested. Figure 6.7 shows the fitted exponential function curve based on the interpolated population data for survey townships. Survey townships are much larger than census tracts and have far fewer observations. It is not surprising that the function is a better fit at the township level (as shown in Figure 6.7) than at the tract level (as shown in Figure 6.6).

6.6 DISCUSSION AND SUMMARY

Based on Table 6.3 and Table 6.5, among the six bivariate monocentric functions (linear, logarithmic, power, exponential, Tanner–Sherratt, and Newling's), the exponential function has the best fit overall. It generates the highest R^2 by linear regressions using both census tract data and survey township data. Only by nonlinear regressions does the Tanner–Sherratt model have a R^2 that is slightly better than the exponential model. Newling's model has two terms (distance and distance squared), and thus its R^2 is not comparable to the bivariate functions. In fact, Newling's model is not a very good fit for the Chicago population density pattern since the term *distance squared* is not statistically significant by linear regressions on both census tract data and survey township data, and is significant by nonlinear regression only at 0.05 based on the survey township data.

We here use the exponential function as an example to compare the regression results by linear and nonlinear regressions. The nonlinear regression yields a slightly

TABLE 6.5
Regressions Based on Monocentric Functions (115 Survey Townships)

Regression Techniques	Functions	a (or A)	b	R-squared
Linear	$D_r = a + br$	328.15	−4.4710	0.4890
	$D_r = a + blnr$	853.13	−198.6594	0.6810
	$lnD_r = A + blnr$	11.55	−2.1165	0.5966
	$lnD_r = A + br$	6.65	−0.0607	0.6952
	$lnD_r = A + br^2$	5.21	−0.0005	0.6292
	$lnD_r = A + b_1r + b_2r^2$	7.02	$b_1 = -0.0780$	0.6989
			$b_2 = 0.0002^a$	
Nonlinear	$D_r = ar^b$	1307.11	−0.6834	0.4587
	$D_r = ae^{br}$	862.74	−0.0574	0.7725
	$D_r = ae^{br^2}$	655.39	−0.0019	0.7770
	$D_r = ae^{b_1r+b_2r^2}$	758.73	$b_1 = 0.0331^b, b_2 = -0.0007^c$	0.7785
Weighted	$D_r = ae^{br}$	807.18	−0.0580	0.8103[d]

[a]Not significant.
[b]Significant at 0.01.
[c]Significant at 0.05 (all other significant at 0.001).
[d]Pseudo-R-squared is not a measure of goodness of fit.

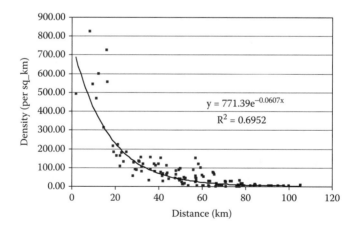

FIGURE 6.7 Density vs. distance exponential trend line (survey townships).

higher R^2 than the linear counterpart using both the tract and township data. The nonlinear regression also generates a higher intercept than the linear regression ($10013 > e^{8.7953} = 6603.1$ for census tracts and $862.74 > e^{6.65} = 771.4$ for survey townships). Recall the argument discussed in Section 6.3 that nonlinear regression tends to value high-density areas more than low-density areas in minimizing the *residual sum squared* (RSS). In other words, the logarithmic transformation in linear regression reduces the error contributions by high-density areas, and thus its fitting intercept tends to swing lower. In weighted regression, observations are weighted by their area sizes. The intercept obtained by weighted regression is between the values by the linear and nonlinear regressions ($6603 < 9157.5 < 10013$ for census tracts and $771.4 < 807.18 < 862.74$ for survey townships). The trend for the slope is unclear: steeper by nonlinear than by linear regression using the census tract data, but flatter by nonlinear than by linear regression using the survey township data.

In weighted regression, observations are weighted by their area sizes. As the area size for survey townships is considered uniform, the weighted regressions are not needed. Indeed, the regression results between weighted and unweighted (regular nonlinear) regression are very similar based on the survey township data (using the exponential function as an example), but the differences are evident based on the census tract data.

Most of the findings discussed above are consistent as the analysis unit changes from census tracts to survey townships. That implies that the MAUP is not a major concern for density function fittings in the study area. Data aggregated at the survey township level smooth out variations in both densities and distances. As expected, R^2 is higher and the intercept is lower for each function obtained from the survey township data than those from the census tract data. However, the change in the slope is not consistent (e.g., in the exponential function, the slope is steeper for survey townships than for census tracts by linear regressions, but flatter for survey townships than for census tracts by nonlinear regressions).

From Table 6.4 the regression results based on assumption 1 reveal that within most of the proximal areas population densities decline with distance from their nearest centers. This is particularly true for suburban centers. But there are areas with a reversed pattern, with densities increasing with distances, particularly in the central city (e.g., centers 14, 0, 1, and 4; refer to Figure 6.5 for their locations). This clearly indicates density craters around these downtown or near-downtown job centers because of significant nonresidential land uses (commercial, industrial, or public) near these centers.[11] The analysis of density function fittings based on polycentric assumption 1 enables us to examine the effect of centers at the local areas surrounding the centers. The regression result based on assumption 2 indicates distance decay effects for most centers: seven of the coefficients b_i have the expected negative signs and are statistically significant, and four more have the expected signs. The regression based on assumption 3 is most difficult to implement, particularly when the number of centers increases. Based on the author's experiences, models with more than a half-dozen centers are hard to converge, and efforts should be made to reduce the number of centers considered by eliminating centers of less significance. The regression result based on assumption 4 indicates that the CBD exerts the dominant effect on the density pattern, and the effects from the nearest subcenters are also significant.

Function fittings are a common tool for many quantitative analysis tasks. We often find ourselves looking for ways to characterize how an activity or event varies with distance from a source. Urban and regional studies suggest that in addition to population density, land use intensity (reflected in its price), land productivity, commodity prices, and wage rate may all experience "distance decay effects" (Wang and Guldmann, 1997), and studies testing the spatial patterns may benefit from the function-fitting approach. Furthermore, various patterns (backwash vs. spread, centralization vs. decentralization) can be identified by examining the changes over time and analyzing whether the changes also vary in direction (e.g., northward vs. southward, along certain transportation routes, etc.).

APPENDIX 6A: DERIVING URBAN DENSITY FUNCTIONS

This appendix discusses the theoretical foundations for urban density functions: an economic model following Mills (1972) and Muth (1969) and a gravity-based model based on Wang and Guldmann (1996).

MILLS–MUTH ECONOMIC MODEL

Assuming urban residents with the same income and preferences, each intends to maximize his or her utility, $U(h,x)$, determined by the amount of land (i.e., housing lot size) consumed, h, and everything else, x. The budget constraint y is given as

$$y = p_h h + p_x x + tr$$

where p_h and p_x are the prices of land and everything else, respectively, t is the unit cost for transportation to the center of the city, and r is the distance to the center.

The utility maximization yields a first-order condition given by

$$\frac{dU}{dr} = 0 = \frac{dp_h}{dr} h + t \tag{A6.1}$$

Assume that the price elasticity of the land demand is –1 (i.e., often referred to as the assumption of "negative unit elasticity for housing demand") such as

$$h = p_h^{-1} \tag{A6.2}$$

Combining Equations A6.1 and A6.2 yields the negative exponential rent gradient:

$$\frac{1}{p_h} \frac{dp_h}{dr} = -t \tag{A6.3}$$

As population density $D(r)$ is given by the inverse of lot size h (i.e., $D(r) = 1/h$), we have $D(r) = 1/(p_h^{-1}) = p_h$. Substituting into Equation A6.3 and solving the differential equation yields the negative exponential density function

$$D(r) = D_0 e^{-tr} \tag{A6.4}$$

Recoding the constant term D_0 as the CBD intercept a and the unit cost of transportation t (plus a negative sign) as the density gradient b, Equation A6.4 becomes $D(r) = ae^{br}$, i.e., Equation 6.1. See Fisch (1991) for details.

GRAVITY-BASED MODEL

Consider a city composed of n equal-area tracts. The population in tract j, x_j, may be expressed as a linear function of the potential there, with

$$kx_j = \sum_{i=1}^{n} \frac{x_i}{d_{ij}^{\beta}} \qquad (A6.5)$$

where d_{ij} is the distance between tract i and j, β is the distance friction coefficient, and n is the total number of tracts in the city. This proposition assumes that population at a particular location is determined by its location advantage or accessibility to all other locations in the city, measured as a gravity potential.

Equation A6.5 can also be written, in matrix notation, as

$$k\mathbf{X} = \mathbf{A}\mathbf{X} \qquad (A6.6)$$

where $\mathbf{X}$ is a column vector of n elements $(x_1, x_2, \ldots, x_n)$, $\mathbf{A}$ is an $n \times n$ matrix with terms involving the d_{ij}'s and β, and k is an unknown scalar. Normalizing one of the population terms, say, $x_1 = 1$, Equation A6.6 becomes a system of n equations with n unknown variables that can be solved by numerical analysis methods.

Assuming a transportation network made of circular and radial roads (see Chapter 11, similar to Figure 11.3) that define the distance term d_{ij}, and picking up a β value, the population distribution pattern can be simulated in the city. The simulation indicates that the density pattern conforms to the negative exponential function when β is within a particular range (i.e., $0.2 \leq \beta < 1.5$). When β is larger (i.e., $1.5 \leq \beta \leq 2.0$), the log-linear function becomes a better fit. Therefore, the model allows for the flexibility of best-fitting function forms in various times and in different cities, as suggested by empirical studies.

This appendix shows that the widely used negative exponential density function can be derived from either the Mills–Muth economic model or the gravity-based model. It is not surprising because of their common ground in characterizing individual behavior: the gravity model itself can be derived from individual utility maximization by an economic model (see Appendix 3).

APPENDIX 6B: OLS REGRESSION FOR A LINEAR BIVARIATE MODEL

A linear bivariate regression model is written as

$$y_i = a + bx_i + e_i$$

where x is a predictor (independent variable) of y (dependent variable), e is the error term or residual, i indexes individual cases or observations, and a and b are parameter estimates (often referred to as intercept and slope, respectively).

Residuals e measure prediction error, i.e.,

$$e_i = y_i - (a + bx_i)$$

When $e > 0$, the actual y is higher than predicted (i.e., underprediction). When $e < 0$, the actual y is lower than predicted (i.e., overprediction). A perfect prediction results in a zero residual ($e = 0$). Since either underprediction or overprediction contributes to inaccuracy, a good measure for overall accuracy of the predictions is the *residual sum squared* (RSS):

$$RSS = \sum_i e_i^2 = \sum_i (y_i - a - bx_i)^2 \tag{A6.7}$$

The *ordinary least squares* (OLS) *regression* seeks the optimal values of a and b so that RSS is minimized. Here x_i and y_i are observed, and a and b are the only variables. The optimization conditions for Equation A6.7 are

$$\partial RSS / \partial a = -2 \sum_i (y_i - a - bx_i) = 0 \tag{A6.8}$$

$$\partial RSS / \partial b = 2 \sum_i (y_i - a - bx_i)(-x_i) = 0 \tag{A6.9}$$

Assuming that n is the total number of observations, we have $\sum_i a = na$. Solve Equation A6.8 for a:

$$a = \frac{1}{n}\sum_i y_i - \frac{b}{n}\sum_i x_i \tag{A6.10}$$

Substituting Equation A6.10 into Equation A6.9 and solving for b yields

$$b = \frac{n\sum_i x_i y_i - \sum_i x_i \sum_i y_i}{n\sum_i x_i^2 - (\sum_i x_i)^2} \tag{A6.11}$$

Substituting Equation A6.11 back into Equation A6.10 yields the solution to a.

APPENDIX 6C: SAMPLE SAS PROGRAM FOR MONOCENTRIC FUNCTION FITTINGS

```
/* Monocent.sas runs linear, nonlinear & weighted
   regressions based on various monocentric models
   by Fahui Wang, 1-31-2005*/

/* Input data */
data mono;
   infile 'c:\gis_quant_book\projects\chicago\monodist.csv'
      dlm=',';
   input popden dist area;
   dist=dist/1000; area=area/1000000; /*convert units to km,
      km_sq */
   lndist=log(dist);
   lnpopden=log(popden+1); /*to avoid taking log of 0 */
   dist_sq=dist**2;

/* the following codes variables used in the Cubic spline model
   by assigning arbitrary x0, x1, x2 &x3*/
   x0=1.0; x1=5.0; x2=10.0; x3=15.0;
   z1=0; z2=0; z3=0;
   if dist > x1 then z1=1;
   if dist > x2 then z2=1;
   if dist > x3 then z3=1;
   v1=dist-x0; v2=(dist-x0)**2; v3=(dist-x0)**3;
   v4=z1*(dist-x1)**3; v5=z2*(dist-x2)**3; v6=z3*(dist-x3)**3;
/* proc means;*/

proc reg; /* simple OLS linear regressions */
   model popden = dist; /*linear model */
   model popden = lndist; /*logarithmic model */
   model lnpopden = lndist; /*power model*/
   model lnpopden = dist; /*exponential model */
   model lnpopden = dist_sq; /*Tanner-Sherratt model */
   model lnpopden = dist dist_sq; /*Newling's model */
   model popden = v1 v2 v3 v4 v5 v6; /*Cubic spline model */
proc model; /* nonlinear regression on power func */
   parm a 30000 b 0.0; /*assign starting values */
   popden = a*dist**b; /*code power function */
   fit popden;
```

```
proc model; /* nonlinear regression on exponential func */
   parm a 30000 b 0.0;
   popden = a*exp(b*dist);
   fit popden;
proc model; /* nonlinear regression on Tanner-Sherratt */
   parm a 30000 b 0.0;
   popden = a*exp(b*dist_sq);
   fit popden;
proc model; /* nonlinear regression on Newling's */
   parm a 30000 b1 0.0 b2 0.0;
   popden = a*exp(b1*dist+b2*dist_sq);
   fit popden;
proc model; /* weighted regression on exponential func */
parm a 30000 b 0.0;
popden = a*exp(b*dist);
fit popden;
weight area;
run;
```

NOTES

1. As a young graduate student in the early 1990s taking an urban economics class taught by Donald Haurin at The Ohio State University, I was fascinated by the regularity of urban population density patterns, and even more so by the elegant economic model that explains the empirical patterns.
2. In the U.S., the TIGER/line files use the CFCC code D65 to identify the location of government center in the point file.
3. The contour map of employment density was based on the logarithm of density. The direct use of density values would lead to too many contour lines in high-density areas.
4. Population density peaks may not qualify as centers, as we have learned from Newling's model for the monocentric structure. Commercial and other business entities often dominate the land use in an employment center, which exhibits a local population density crater.
5. These are job centers and do not necessarily have the highest population densities (see Figure 6.5). The classic urban economic model for explaining urban density functions assumes a single center at the CBD, where all jobs are located. Polycentric models extend the assumption to multiple job centers in a city. Therefore, centers should be based on the employment instead of population distribution pattern. For cities in the U.S., the data source for employment location is a special census data: CTPP (Census Transportation Planning Package). See Section 5.5.
6. In Section 6.5.2, step 1, the tool "Near" will be executed again, and both fields NEAR_FID and NEAR_DIST will be updated to represent the nearest centers and distances from them. The field DIST_CBD is added here to preserve the distances from the CBD.

7. For example, use the formula = LN( ) for natural logarithms in Excel.
8. The choice of adding 1 (instead of 0.2, 0.5, or others) is arbitrary and may bias the coefficient estimates. However, different additive constants have minimal consequence for significance testing, as standard errors grow proportionally with the coefficients and thus leave the t values unchanged (Osgood, 2000, p. 36).
9. One may also fit the cubic spline function in Excel. For example, by assigning arbitrary values to the parameters x_i, such as $x_0 = 1, x_1 = 5, x_2 = 10, x_3 = 15$, the cubic spline model has a $R^2 = 0.43$, but most of the terms are not statistically significant.
10. Note that the "Near" tool is already executed once in step 3 of Part 1, and the fields NEAR_FID and NEAR_DIST will be updated.
11. For example, fitting Newling's model in the proximal area (489 tracts) around the CBD (center 14) by linear regression yields

$$lnD = 7.7076 + 0.1970\ r_{14} - 0.0094\ r_{14}^2,\ \text{with}\ R^2 = 0.03.$$
$$(32.45)\quad (3.18)\qquad\ (-2.48)$$

Despite the low R^2, the t values in parentheses indicate that both the terms r_{14} and r_{14}^2 are statistically significant, and the positive coefficient for r_{14} and the negative coefficient for r_{14}^2 confirm the existence of density crater at the *local* level. The model by nonlinear regression yields similar results.

7 Principal Components, Factor, and Cluster Analyses, and Application in Social Area Analysis

This chapter discusses three important multivariate statistical analysis methods: principal components analysis (PCA), factor analysis (FA), and cluster analysis (CA). PCA and FA are often used together for data reduction by structuring many variables into a limited number of components (factors). The techniques are particularly useful for eliminating variable collinearity and uncovering latent variables. Applications of the methods are widely seen in socioeconomic studies (also see case study 8 in Section 8.4). While the PCA and FA group variables, the CA classifies many observations into categories according to similarity among their attributes. In other words, given a dataset as a table, the PCA and FA reduce the number of columns and the CA reduces the number of rows.

Social area analysis is used to illustrate the techniques, as it employs all three methods. The interpretation of social area analysis results also leads to a review and comparison of three classic models on urban structure, namely, the concentric zone model, the sector model, and the multinuclei model. The analysis demonstrates how analytical statistical methods synthesize descriptive models into one framework. Beijing, the capital city of China, on the verge of forming its social areas after decades under a socialist regime, is chosen as the study area for a case study. Usage of GIS in this case study is limited to mapping for spatial patterns.

Section 7.1 discusses principal components and factor analysis. Section 7.2 explains cluster analysis. Section 7.3 reviews social area analysis. A case study on the social space in Beijing is presented in Section 7.4 to provide a new perspective to the fast-changing urban structure in China. The chapter is concluded with a discussion and brief summary in Section 7.5.

7.1 PRINCIPAL COMPONENTS AND FACTOR ANALYSIS

Principal components and factor analysis are often used together for data reduction. Benefits of this approach include uncovering latent variables for easy interpretation and removing multicollinearity for subsequent regression analysis. In many socioeconomic applications, variables extracted from census data are often correlated with each other, and thus contain duplicated information to some extent. Principal components and factor analysis use fewer factors to represent the original variables, and thus simplify the structure for analysis. Resulting component or factor scores are uncorrelated to each other (if not rotated or orthogonally rotated), and thus can be used as explanatory variables in regression analysis.

Despite the commonalities, principal components and factor analysis are "both conceptually and mathematically very different" (Bailey and Gatrell, 1995, p. 225). Principal components analysis uses the same number of variables (components) to simply transform the original data, and thus is a mathematical transformation (strictly speaking, not a statistical operation). Factor analysis uses fewer variables (factors) to capture most of the variation among the original variables (with error terms), and thus is a statistical analysis process. Principal components attempts to explain the variance of observed variables, whereas factor analysis intends to explain their intercorrelations (Hamilton, 1992, p. 252). In many applications (as in ours), the two methods are used together. In SAS, principal components analysis is offered as an option under the procedure for factor analysis.

7.1.1 PRINCIPAL COMPONENTS FACTOR MODEL

In formula, *principal components analysis* (PCA) transforms original data on K observed variables Z_k to data on K principal components F_k that are independent from (uncorrelated with) each other:

$$Z_k = l_{k1}F_1 + l_{k2}F_2 + ... + l_{kj}F_j + ... + l_{kK}F_K \tag{7.1}$$

Retaining only the J largest components ($J < K$), we have

$$Z_k = l_{k1}F_1 + l_{k2}F_2 + ... + l_{kJ}F_J + v_k \tag{7.2}$$

where the discarded components are represented by the residual term v_k, such as

$$v_k = l_{k,J+1}F_{J+1} + l_{k,J+2}F_{J+2} + ... + l_{kK}F_K \tag{7.3}$$

Equations 7.2 and 7.3 represent a model termed *principal components factor analysis* (PCFA). The PCFA retains the largest components to capture most of the variance while discarding minor components with small variance. The PCFA is the method used in social area analysis (Cadwallader, 1996, p. 137) and is simply referred to as factor analysis in the remainder of this chapter.

In a true *factor analysis* (FA), the residual (error) term, denoted as u_k to distinguish it from v_k in a PCFA, is unique to each variable Z_k:

$$Z_k = l_{k1}F_1 + l_{k2}F_2 + ... + l_{kJ}F_J + u_k$$

The u_k are termed *unique factors* (in contrast to *common factors* F_j). In the PCFA, the residual v_k is a linear combination of the discarded components ($F_{J+1}, ..., F_K$) and thus cannot be uncorrelated like the u_k in a true FA (Hamilton, 1992, p. 252).

7.1.2 FACTOR LOADINGS, FACTOR SCORES, AND EIGENVALUES

For convenience, the original data of observed variables Z_k are first *standardized[1]* prior to the PCA and FA analysis, and the initial values for components (factors) are also standardized. When both Z_k and F_j are standardized, the l_{kj} in Equations 7.1 and 7.2 are *standardized coefficients* in the regression of variables Z_k on components (factors) F_j, also termed *factor loadings*. For example, l_{k1} is the loading of variables Z_k on standardized component F_1. Factor loading reflects the strength of relations between variables and components.

Conversely, the components F_j can be reexpressed as a linear combination of the original variables Z_k:

$$F_j = a_{1j}Z_1 + a_{2j}Z_2 + ... + a_{Kj}Z_K \qquad (7.4)$$

Estimates of these components (factors) are termed *factor scores*. Estimates of a_{kj} are *factor score coefficients*, i.e., coefficients in the regression of factors on variables.

The components F_j are constructed to be uncorrelated with each other and are ordered such that the first component F_1 has the largest sample variance (λ_1), F_2 the second largest, and so on. The variances λ_j corresponding to various components are termed *eigenvalues*, and $\lambda_1 > \lambda_2 > ....$

Since standardized variables have variances of 1, the total variance of all variables also equals the number of variables, such as

$$\lambda_1 + \lambda_2 + ... + \lambda_K = K \qquad (7.5)$$

Therefore, the *proportion* of total variance explained by the jth component is λ_j/K.

Eigenvalues provide a basis for judging which components (factors) are important and which are not, and thus deciding how many components to retain. One may also follow a rule of thumb that only eigenvalues greater than 1 are important (Griffith and Amrhein, 1997, p. 169). Since the variance of each standardized variable is 1, a component with $\lambda < 1$ accounts for less than an original variable's variation, and thus does not serve the purpose of data reduction.

The eigenvalue-1 rule is arbitrary. A *scree graph* plots eigenvalues against component (factor) number and provides a more useful guidance (Hamilton, 1992, p. 258). For example, Figure 7.1 shows the scree graph of eigenvalues in a case of 14 components (using the result from case study 7 in Section 7.4). The graph levels off after component 4, indicating that components 5 to 14 account for relatively little additional variance. Therefore, four components may be retained as principal components.

Outputs from statistical analysis software such as SAS include important information, such as factor loadings, eigenvalues, and proportions (of total variance). Factor scores can be saved in a predefined external file. The factor analysis procedure in SAS also outputs a correlation matrix between the observed variables for analysts to examine their relations.

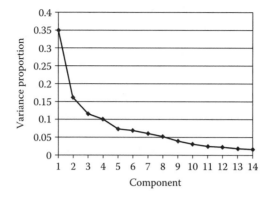

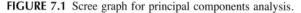

FIGURE 7.1 Scree graph for principal components analysis.

7.1.3 ROTATION

Initial results from PCFA are often hard to interpret as variables load across factors. While fitting the data equally well, *rotation* generates simpler structure and more interpretable factors by maximizing the loading (positive or negative) of each variable on one factor and minimizing the loadings on the others. As a result, we can detect which factor (latent variable) captures the information contained in what variables (observed), and subsequently label the factors adequately.

Orthogonal rotation generates independent (uncorrelated) factors, an important property for many applications. A widely used orthogonal rotation method is *Varimax rotation*, which maximizes the variance of the squared loadings for each factor, and thus polarizes loadings (either high or low on factors). Varimax rotation is often the rotation technique used in social area analysis. *Oblique rotation* (e.g., *promax rotation*) generates even greater polarization, but allows correlation between factors. In SAS, an option is provided to specify which rotation to use.

As a summary, Figure 7.2 illustrates the process of PCFA:

1. The original dataset of K observed variables with n records is first standardized to a dataset of Z scores with the same number of variables and records.
2. PCA then uses K uncorrelated components to explain all the variance of the K variables.
3. PCFA keeps only J ($J < K$) principal components to capture most of the variance.
4. A rotation method is used to load each variable strongly on one factor (and near zero on the others) for easier interpretation.

The SAS procedure for factor analysis (FA) is FACTOR, which also reports the principal components analysis (PCA) results preceding those of FA. The following sample SAS statements implement the factor analysis that uses four factors to capture the structure of 14 variables, x1 through x14, and adopts the Varimax rotation technique:

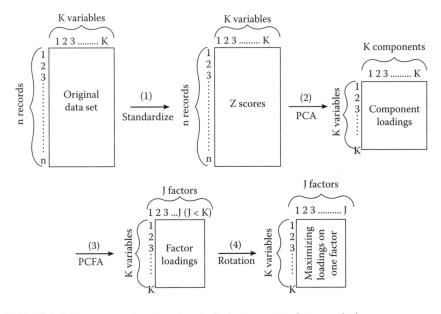

FIGURE 7.2 Data processing steps in principal components factor analysis.

```
proc factor out=FACTSCORE (replace=yes)

    nfact=4 rotate=varimax;

    var x1-x14;
```

The SAS data set FACTSCORE has the factor scores, which can be saved to an external file. Note that a SAS program is not case sensitive.

7.2 CLUSTER ANALYSIS

Cluster analysis (CA) groups observations according to similarity among their attributes. As a result, the observations within a cluster are more similar than observations between clusters, as measured by the clustering criterion. Note the difference between CA and another similar multivariate analysis technique — discriminant function analysis (DFA). Both group observations into categories based on the characteristic variables. Categories are unknown in CA but known in DFA. See Appendix 7A for further discussion on DFA.

Geographers have a long-standing interest in cluster analysis (CA) that has been developed in applications such as regionalization and city classification. In the case of social area analysis, cluster analysis is used to further analyze the results from factor analysis (i.e., factor scores of various components across space) and group areas into different types of social areas.

A key element in deciding assignment of observations to clusters is *distance*, measured in various ways. The most commonly used distance measure is Euclidean distance:

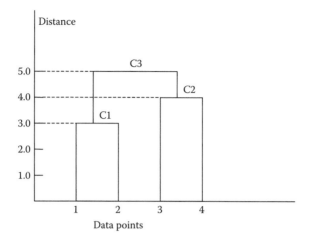

FIGURE 7.3 Dendrogram for the clustering analysis example.

$$d_{ij} = (\sum_{k=1}^{K} (x_{ik} - x_{jk})^2)^{1/2} \tag{7.6}$$

where x_{ik} and x_{jk} are the kth variable value of the K-dimensional observations for individuals i and j. When $K = 2$, Euclidean distance is simply the straight-line distance between observations i and j in a two-dimensional space. Like the various distance measures discussed in Chapter 2, distance measures here also include Manhattan (or city block) distance and others (e.g., Minkowski distance, Canberra distance) (Everitt et al., 2001, p. 40).

The most widely used clustering method is the *agglomerative hierarchical methods* (AHMs). The methods produce a series of groupings: the first consists of single-member clusters, and the last consists of a single cluster of all members. The results of these algorithms can be summarized with a *dendrogram*, a tree diagram showing the history of sequential grouping process. See Figure 7.3 for the example illustrated below. In the diagram, the clusters are nested and each cluster is a member of a larger, higher-level cluster.

For illustration, an example is used to explain a simple AHM, the *single-linkage method* or the *nearest-neighbor method*. Consider a dataset of four observations with the following distance matrix:

$$D_1 = \begin{matrix} 1 \\ 2 \\ 3 \\ 4 \end{matrix} \begin{bmatrix} 0 & & & \\ 3 & 0 & & \\ 6 & 5 & 0 & \\ 9 & 7 & 4 & 0 \end{bmatrix}$$

The smallest no-zero entry in the above matrix D_1 is $(2 \rightarrow 1) = 3$, and therefore individuals 1 and 2 are grouped together to form the first cluster C1. Distances between this cluster and the other two individuals are defined according to the nearest-neighbor criterion:

$$d_{(12)3} = \min\{d_{13}, d_{23}\} = d_{23} = 5$$

$$d_{(12)4} = \min\{d_{14}, d_{24}\} = d_{24} = 7$$

A new matrix is now obtained with cells representing distances between cluster C1 and individuals 3 and 4, or between individuals 3 and 4:

$$D_2 = \begin{array}{c} (12) \\ 3 \\ 4 \end{array} \begin{bmatrix} 0 & & \\ 5 & 0 & \\ 7 & 4 & 0 \end{bmatrix}$$

The smallest no-zero entry in D_2 is $(4 \rightarrow 3) = 4$, and thus individuals 3 and 4 are grouped to form a cluster C2. Finally, clusters C1 and C2 are grouped together, with distance equal to 5, to form one cluster C3 containing all four members. The process is summarized in a dendrogram in Figure 7.3, where the height represents the distance at which each fusion is made.

Similarly, the *complete linkage (farthest-neighbor) method* uses the maximum distance between pair of objects (one in one cluster and one in the other); the *average linkage method* uses the average distance between pair of objects; and the *centroid method* uses squared Euclidean distance between individuals and cluster means (centroids).

Another commonly used AHM is *Ward's method*. The objective at each stage is to minimize the increase in the total within-cluster error sum of squares given by

$$E = \sum_{c=1}^{C} E_c$$

where

$$E_c = \sum_{i}^{n_c} \sum_{k=1}^{K} \left(x_{ck,j} - \overline{x_{ck}}\right)^2$$

in which $x_{ck,i}$ is the value for the kth variable for the ith observation in the cth cluster, and $\overline{x_{ck}}$ is the mean of the kth variable in the cth cluster.

Each clustering method has its advantages and disadvantages. A desirable clustering should produce clusters of similar size, densely located, compact in shape, and internally homogeneous (Griffith and Amrhein, 1997, p. 217). The single-linkage method tends to produce unbalanced and straggly clusters and should be avoided in most cases. If outlier is a major concern, the centroid method should be used. If compactness of clusters is a primary objective, the complete linkage method should be used. Ward's method tends to find same size and spherical clusters and is recommended if no single overriding property is desired (Griffith and Amrhein, 1997, p. 220). The case study in this chapter also uses Ward's method.

The choice for the number of clusters depends on objectives of specific applications. Similar to the selection of factors based on the eigenvalues in factor analysis, one may also use a scree plot to assist in the decision. In the case of Ward's method, a graph of R^2 vs. the number of clusters helps choose the number, beyond which little more homogeneity is attained by further mergers.

In SAS, the procedure CLUSTER implements the cluster analysis and the procedure TREE generates the dendrogram. The following sample SAS statements use Ward's method for clustering and cut off the dendrogram at nine clusters:

```
proc cluster method=ward outtree=tree;

    id subdist_id; /* variable for labeling ids */

var factor1-factor4; /* variables used */

  proc tree out=bjcluster ncl=9;

    id subdist_id;
```

7.3 SOCIAL AREA ANALYSIS

The social area analysis was developed by Shevky and Williams (1949) in a study of Los Angeles and was later elaborated on by Shevky and Bell (1955) in a study of San Francisco. The basic thesis is that the changing social differentiation of society leads to residential differentiation within cities. The studies classified census tracts into types of social areas based on three basic constructs: economic status (social rank), family status (urbanization), and segregation (ethnic status). Originally the three constructs were measured by six variables: economic status was captured by occupation and education; family status by fertility, women labor participation, and single-family houses; and ethnic status by percentage of minorities (Cadwallader, 1996, p. 135). In a factor analysis, an idealized factor loadings matrix probably looks like Table 7.1. Subsequent studies using a large number and variety of measures generally confirmed the validity of the three constructs (Berry, 1972, p. 285; Hartshorn, 1992, p. 235).

Geographers made an important advancement in social area analysis by analyzing the spatial patterns associated with these dimensions (e.g., Rees, 1970; Knox, 1987). The socioeconomic status factor tends to exhibit a sector pattern: tracts with high values for variables, such as income and education, form one or more sectors, and low-status tracts form other sectors. The family status factor tends to form concentric zones: inner zones are dominated by tracts with small families with either very young or very old household heads, and tracts in outer zones are mostly

TABLE 7.1
Idealized Factor Loadings in Social Area Analysis

	Economic Status	Family Status	Ethnic Status
Occupation	I	O	O
Education	I	O	O
Fertility	O	I	O
Female labor participation	O	I	O
Single-family house	O	I	O
Minorities	O	O	I

Note: I denotes a number close to 1 or –1; O denotes a number close to 0.

occupied by large families with middle-age household heads. The ethnic status factor tends to form clusters, each of which is dominated by a particular ethnic group. Superimposing the three constructs generates a complex urban mosaic, which can be grouped into various social areas by cluster analysis. See Figure 7.4. By studying the spatial patterns from social area analysis, three classic models for urban structure — Burgess's (1925) concentric zone model, Hoyt's (1939) sector model, and the Ullman–Harris (Harris and Ullman, 1945) multinuclei model — are synthesized into one framework. In other words, each of the three models reflects one specific dimension of urban structure and is complementary to the others.

There are at least three criticisms of the factorial ecological approach to understanding residential differentiation in cities (Cadwallader, 1996, p. 151). First, the analysis results are sensitive to research design, such as variables selected and measured, analysis units, and factor analysis methods. Second, it is still a descriptive form of analysis and fails to explain the underlying process that causes the patterns. Third, the social areas identified by the studies are merely homogeneous, but not necessarily functional regions or cohesive communities. Despite the criticisms, social area analysis helps us understand residential differentiation within cities, and serves as an important instrument for studying intraurban social spatial structure. Applications of social area analysis can be seen on cities in developed countries, particularly rich on cities in North America (see a review by Davies and Herbert, 1993), and also on some cities in developing countries (e.g., Berry and Rees, 1969; Abu-Lughod, 1969).

7.4 CASE STUDY 7: SOCIAL AREA ANALYSIS IN BEIJING

This case study is developed on the basis of a research project reported in Gu et al. (2005). Detailed research design and interpretation of the results can be found in the original paper. This section shows the procedures to implement the study, with emphasis on illustrating the three statistical methods. In addition, the study illustrates how to test the spatial structure of factors by regression models with dummy variables. Since the 1978 economic reforms in China, and particularly the 1984 urban reforms, including the urban land use reform and the housing reform, urban landscape in China has changed significantly. Many large cities have been on the

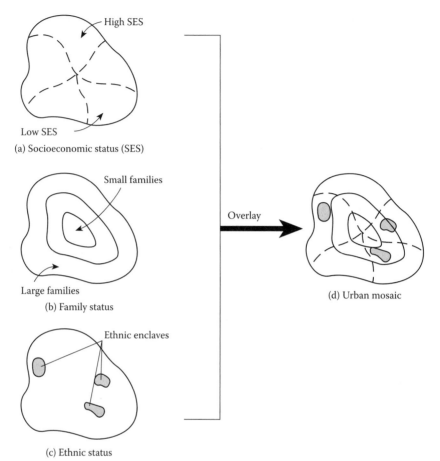

FIGURE 7.4 Conceptual model for urban mosaic.

transition from self-contained work unit neighborhood systems to more differentiated urban space. As the capital city of China, Beijing offers an interesting case to look into this important change in urban structure in China.

The study area is the contiguous urbanized area of Beijing, with 107 subdistricts (*jiedao*), excluding the 2 remote suburban districts (Mentougou and Fangshan) and 23 subdistricts on the periphery of inner suburbs (also rural and lack of complete data). See Figure 7.5. The study area had a total population of 5.9 million, and the subdistricts had an average population of 55,200 in 1998. Subdistrict has been the basic administrative unit in Beijing for decades, and also the lowest geographic level reported in government statistical reports accessible by the public. Therefore, it was the analysis unit used in this research. Because of the lack of socioeconomic data in the national census of population, most of the data used in this research were extracted from the 1998 statistical yearbooks of individual districts in Beijing. Some data, such as personal income and individual living space, were obtained through a survey of households conducted in 1998.

FIGURE 7.5 Study area for Beijing's social area analysis.

The following datasets are provided:

1. Shapefile `bjsa` contains 107 urban subdistricts (*jiedao*) in Beijing.
2. Text file `bjattr.csv` is the attribute dataset.

In the shapefile `bjsa`, the field `sector` identifies four sectors (= 1 for NE, 2 for SE, 3 for SW, and 4 for NW) and the field `ring` identifies four zones (= 1 for the most inner zone, 2 for the next inner zone, and so on). Sectors and zones are needed for testing spatial structures of social space. Both the shapefile `bjsa` and the attribute file `bjattr.csv` contain a field `ref_id` (subdistrict IDs) as the common key for joining them together. The text file `bjattr.csv` has 14 socio-economic variables (X1 to X14) for social area analysis. Variable names and their basic statistics for these 14 variables are summarized in Table 7.2.

1. *Executing principal components factor analysis in SAS*: See the sample SAS program `FA_Clust.sas` in Appendix 7B (also provided on the CD). The first part of the program uses the SAS procedure PROC FACTOR to

TABLE 7.2
Basic Statistics for Socioeconomic Variables in Beijing (n = 107)

Index	Variables	Mean	Std. Dev.	Minimum	Maximum
X1	Population density (persons/km^2)[a]	14,797.09	13,692.93	245.86	56,378.00
X2	Natural growth rate (%)	−1.11	2.79	−16.41	8.58
X3	Sex ratio (M/F)	1.03	0.08	0.72	1.32
X4	Labor participation ratio (%)[b]	0.60	0.06	0.47	0.73
X5	Household size (persons/household)	2.98	0.53	2.02	6.55
X6	Dependency ratio[c]	1.53	0.22	1.34	2.14
X7	Income (yuan/person)	29,446.49	127,223.03	7505.00	984,566.00
X8	Public service density (no./km^2)[d]	8.35	8.60	0.05	29.38
X9	Industry density (no./km^2)	1.66	1.81	0.00	10.71
X10	Office/retail density (no./km^2)	14.90	15.94	0.26	87.86
X11	Ethnic enclave (0, 1)[e]	0.10	0.31	0.00	1.00
X12	Floating population ratio (%)[a]	6.81	7.55	0.00	65.59
X13	Living space (m^2/person)	8.89	1.71	7.53	15.10
X14	Housing price (yuan/m^2)	6686.54	3361.22	1400.00	18,000.00

[a]The household registration (*hukou*) system in China classifies residents into two categories: permanent and temporal residents. Temporal residents were those migrants from rural areas who had not obtained the permanent residence status in a city, also called floating population. Population density here is measured as number of permanent residents per square kilometer.

[b]Labor participation ratio is percentage of persons in the labor force out of the total population at the working ages, i.e., 18 to 60 years old for males and 18 to 55 years old for females.

[c]Dependency ratio is the number of dependents divided by the number of persons in the labor force.

[d]Public service density is the number of governmental agencies, nonprofit organizations, educational units, hospitals, and postal and telecommunication units per square kilometer.

[e]Ethnic enclave is a dummy variable that identifies whether a minority (mainly Muslims in Beijing) or migrant concentrated area was present in a subdistrict.

execute the principal components factor analysis (PCFA). The program reads the attribute dataset `bjattr.csv`, and uses four factors to capture most of the information in the original 14 variables (x1, x2, ..., x14). The output factor scores are saved in a text file `factscore.csv`, containing the original 14 variables and the 4 factor scores.

The SAS procedure FACTOR reports the result of principal components analysis (PCA) preceding that of factor analysis (FA). The number of factors (four, in our case) is determined by analyzing the eigenvalues from PCA (see Table 7.3).[2] In deciding the number of components (factors) to include in the next step of FA, one has to make a trade-off between the total variance explained (higher by including more components) and inter-pretability of factors (better with less components). Following the eigen-value-1 rule (see Section 7.1.2), we retain four factors. The four components explain over 70% of the total variance. The choice of retaining four components is also supported by analyzing the scree graph in Figure 7.1.

TABLE 7.3
Eigenvalues from Principal Components Analysis

Component	Eigenvalue	Proportion	Cumulative
1	4.9231	*0.3516*	0.3516
2	2.1595	*0.1542*	0.5059
3	1.4799	*0.1057*	0.6116
4	1.2904	*0.0922*	0.7038
5	0.8823	0.0630	0.7668
6	0.8286	0.0592	0.8260
7	0.6929	0.0495	0.8755
8	0.5903	0.0422	0.9176
9	0.3996	0.0285	0.9462
10	0.2742	0.0196	0.9658
11	0.1681	0.0120	0.9778
12	0.1472	0.0105	0.9883
13	0.1033	0.0074	0.9957
14	0.0608	0.0043	1.0000

TABLE 7.4
Factor Loadings in Social Area Analysis

Variables	Land Use Intensity	Neighborhood Dynamics	Socioeconomic Status	Ethnicity
Public service density	0.8887	0.0467	0.1808	0.0574
Population density	0.8624	0.0269	0.3518	0.0855
Labor participation ratio	−0.8557	0.2909	0.1711	0.1058
Office/retail density	0.8088	−0.0068	0.3987	0.2552
Housing price	0.7433	−0.0598	0.1786	−0.1815
Dependency ratio	0.7100	0.1622	−0.4873	−0.2780
Household size	0.0410	0.9008	−0.0501	0.0931
Floating population ratio	0.0447	0.8879	0.0238	−0.1441
Living space	−0.5231	0.6230	−0.0529	0.0275
Income	0.1010	0.1400	0.7109	−0.1189
Natural growth rate	−0.2550	0.2566	−0.6271	0.1390
Ethnic enclave	0.0030	−0.1039	−0.1263	0.6324
Sex ratio	−0.2178	0.2316	−0.1592	0.5959
Industry density	0.4379	−0.1433	0.3081	0.5815

The Varimax rotation technique is used to maximize the loading of a variable on one factor and minimize the loadings on all others. Table 7.4 presents the rotated factor structure (variables are reordered to highlight the factor loading structure). The four factors are labeled to reflect major variables loaded to each factor:

a. "Land use intensity" is by far the most important factor, explaining 35.16% of the total variance and capturing mainly six variables: three density measures (population density, public service density, and office and retail density), housing price, and two demographic variables (labor participation ratio and dependency ratio).

b. "Neighborhood dynamics" accounts for 15.42% of the total variance and includes three variables: floating population ratio, household size, and living space.

c. "Socioeconomic status" accounts for 10.42% of the total variance and includes two variables: average annual income per capita and population natural growth rate.

d. "Ethnicity" accounts for 9.22% of the total variance and includes three variables: ethnic enclave, sex ratio, and industry density.

2. *Executing cluster analysis in SAS*: The second part of the program FA_Clust.sas uses the SAS procedure PROC CLUSTER to execute the cluster analysis (CA) and produces a complete dendrogram tree of clustering. The procedure PROC TREE uses the option NCL=5 to define the number of clusters, based on which the dendrogram tree is cut off. The program saves the result in a text file cluster5.csv (rename the field name cluster to cluster5 for clarification). Repeat the cluster analysis by changing the option to NCL=9, and save the result to cluster9.csv (rename the field name cluster to cluster9 for clarification).

   The study begins with five basic clusters and expands to nine clusters revealing more detailed spatial patterns. For instance, cluster 2 identified in the five-cluster scenario is further divided to clusters 2, 4, and 5 in the nine-cluster scenario. Each cluster represents a social area.

3. *Mapping factor patterns in ArcGIS*: Open the shapefile bjsa in ArcGIS and join the text file factscore.csv to it based on the common key ref_id. Map the field factor1 (factor score for "land use intensity") as shown in Figure 7.6a, factor2 (factor score for "neighborhood dynamics") as in Figure 7.6b, factor3 (factor score for "socioeconomic status") as in Figure 7.6c, and factor4 (factor score for "ethnicity") as in Figure 7.6d.

4. *Mapping social areas in ArcGIS*: Similar to step 3, join both cluster9.csv and cluster5.csv to the shapefile bjsa in ArcGIS, and map the social areas as shown in Figure 7.7. The five basic social areas are shown in different area patterns, and the nine detailed social areas are identified by their cluster numbers.

   For understanding the characteristics of each social area, use the Summarize tool on the merged table in ArcGIS to compute the mean values of factor scores within each cluster. The results are reported in Table 7.5. The clusters are labeled by analyzing the factor scores and the locations relative to the city center.

5. *Testing spatial structure by regressions with dummy variables*: Regression models can be constructed to test whether the spatial pattern of a

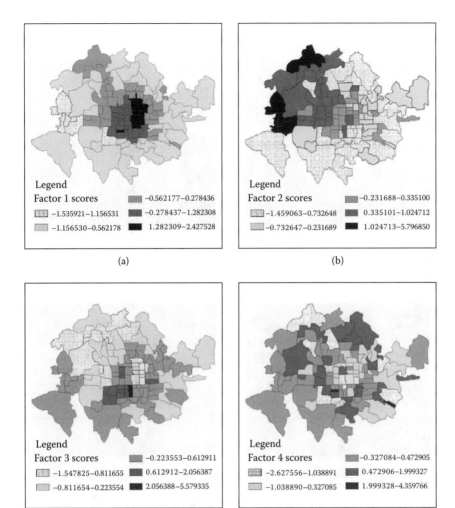

FIGURE 7.6 Spatial patterns of factor scores.

factor is better characterized as zones or sectors (Cadwallader, 1981). Based on the circular ring roads, Beijing is divided into four zones, coded by three *dummy variables* (x_2, x_3, and x_4). Similarly, three additional dummy variables (y_2, y_3, and y_4) are used to code the four sectors (NE, SE, SW, and NW). Table 7.6 shows how the zones and sectors are coded by the dummy variables.

A simple linear regression model for testing the zonal structure can be written as

$$F_i = b_1 + b_2 x_2 + b_3 x_3 + b_4 x_4, \tag{7.7}$$

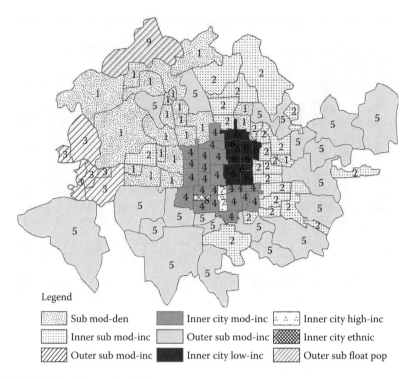

Legend

Sub mod-den	Inner city mod-inc	Inner city high-inc	
Inner sub mod-inc	Outer sub mod-inc	Inner city ethnic	
Outer sub mod-inc	Inner city low-inc	Outer sub float pop	

FIGURE 7.7 Social areas in Beijing.

TABLE 7.5
Characteristics of Social Areas (Clusters)

Clusters Index		Averages of Factor Scores				
Five Clusters	Nine Clusters	No. of Subdistricts	Land Use Intensity	Neighbor-hood Dynamics	Socio-economic Status	Ethnicity
1	1. Suburban moderate density	21	−0.2060	0.6730	−0.6932	0.3583
2	2. Inner suburban moderate income	23	−0.4921	−0.5159	−0.0522	0.4143
	4. Inner city moderate income	22	0.8787	−0.1912	0.5541	0.1722
	5. Outer suburban moderate income	21	−0.8928	−0.8811	0.0449	−0.7247
3	3. Outer suburban manufacturing with high floating population	6	−1.4866	2.0667	0.3611	0.1847
	9. Outer suburban with highest floating population	1	0.1041	5.7968	−0.2505	−1.8765
4	7. Inner city high income	2	0.7168	0.9615	5.1510	−0.8112
	8. Inner city ethnic enclave	1	1.8731	−0.0147	1.8304	4.3598
5	6. Inner city low income	10	2.0570	0.0335	−1.1423	−0.7591

TABLE 7.6
Zones and Sectors Coded by Dummy Variables

Zones		Sectors	
Index and Location	**Codes**	**Index and Location**	**Codes**
1. Inside second ring	$x_2 = x_3 = x_4 = 0$	1. NE	$y_2 = y_3 = y_4 = 0$
2. Between second and third rings	$x_2 = 1, x_3 = x_4 = 0$	2. SE	$y_2 = 1, y_3 = y_4 = 0$
3. Between third and fourth rings	$x_3 = 1, x_2 = x_4 = 0$	3. SW	$y_3 = 1, y_2 = y_4 = 0$
4. Outside fourth ring	$x_4 = 1, x_2 = x_3 = 0$	4. NW	$y_4 = 1, y_2 = y_3 = 0$

where F_i is the score of a factor ($i = 1, 2, 3$, and 4), the constant term b_1 is the average factor score in zone 1 (when $x_2 = x_3 = x_4 = 0$, also referred to as *reference zone*), and the coefficient b_2, b_3, or b_4 is the difference of average factor scores between zone 1 and zone 2, 3, or 4, respectively. Similarly, a regression model for testing the sectoral structure can be written as

$$F_i = c_1 + c_2 y_2 + c_3 y_3 + c_4 y_4, \tag{7.8}$$

where notations have interpretations similar to those in Equation 7.7. Export the attribute table of shapefile `bjsa` (joined with `factscore.csv`) to an external file `zone_sect.dbf` containing the factor scores, and the fields `ring` (identifying zones) and `sector` (identifying sectors). In the file `zone_sect.dbf`, use Excel or SAS to create and compute the dummy variables x_2, x_3, x_4, y_2, y_3, and y_4 according to Table 7.6, and run regression models in Equations 7.7 and 7.8. The results are presented in Table 7.7. A sample SAS program `BJreg.sas` is provided in the CD for reference.

7.5 DISCUSSION AND SUMMARY

R^2 in Table 7.7 indicates whether the zonal or sectoral model is a good fit, and an individual t statistic (in parenthesis) indicates whether a coefficient is statistically significant (i.e., whether a zone or a sector is significantly different from the reference zone or sector). Clearly, the land use intensity pattern fits the zonal model well, and the negative coefficients b_2, b_3, and b_4 are all statistically significant and indicate declining land use intensity from inner to outer zones. The neighborhood dynamics pattern is better characterized by the sectoral model, and the positive coefficient c_4 (statistically significant) confirms high portions of floating population in the northwest sector of Beijing. The socioeconomic status factor displays both zonal and sectoral patterns, but a stronger sectoral structure. The negative coefficients b_3 and b_4 (statistically significant) in the socioeconomic status model imply that factor scores decline toward the third and fourth zones, and the positive coefficient c_3 (statistically

TABLE 7.7
Regressions for Testing Zonal vs. Sectoral Structures ($n = 107$)

Factors		Land Use Intensity	Neighborhood Dynamics	Socioeconomic Status	Ethnicity
Zonal model	b_1	1.2980***	−0.1365	0.4861**	−0.0992
		(12.07)	(−0.72)	(2.63)	(−0.51)
	b_2	−1.2145***	0.0512	−0.4089	0.1522
		(−7.98)	(0.19)	(−1.57)	(0.56)
	b_3	−1.8009***	−0.0223	−0.8408**	−0.0308
		(−11.61)	(−0.08)	(−3.16)	(−0.11)
	b_4	−2.1810***	0.4923	−0.7125**	0.2596
		(−14.47)	(1.84)	(−2.75)	(0.96)
	R^2	0.697	0.046	0.105	0.014
Sectoral model	c_1	0.1929	−0.3803**	−0.3833**	−0.2206
		(1.14)	(−2.88)	(−2.70)	(−1.32)
	c_2	−0.1763	−0.3511	0.4990*	0.6029*
		(−0.59)	(−1.52)	(2.01)	(2.06)
	c_3	−0.2553	0.0212	1.6074***	0.4609
		(−0.86)	(0.09)	(6.47)	(1.58)
	c_4	−0.3499	1.2184***	0.1369	0.1452
		(−1.49)	(6.65)	(0.69)	(0.63)
	R^2	0.022	0.406	0.313	0.051

Note: *, Significant at 0.05; **, significant at 0.01; and ***, significant at 0.001.

significant) indicates higher factor scores in the southwest sector, mainly because of two high-income subdistricts in Xuanwu District. The ethnicity factor does not conform to either the zonal or sectoral model. Ethnic enclaves scatter citywide and may be best characterized by a multiple nuclei model.

Land use intensity is clearly the primary factor forming the concentric social spatial structure in Beijing. From the inner city (clusters 4, 6, 8, and 9) to inner suburbs (clusters 1 and 2) and to remote suburbs (clusters 3, 5, and 7), population densities as well as densities of public services, offices, and retails declined along with land prices. The neighborhood dynamics, mainly the influence of floating population, is the second factor shaping the formation of social areas. Migrants are attracted to economic opportunities in the fast-growing Haidian District (cluster 1) and manufacturing jobs in the Shijingshan District (cluster 3). The effects of the third factor (socioeconomic status) can be found in the emergence of the high-income areas in two inner city subdistricts (cluster 8), and the differentiation between middle-income (cluster 1) and low-income areas (clusters 2, 3, and 5) in suburbs. The fourth factor of ethnicity does not come to play until the number of clusters is expanded to nine.

In Western cities, the socioeconomic status construct plays a dominant force in forming a sectoral pattern, along with the family structure construct featuring a zonal pattern and the ethnicity construct exhibiting a multinuclei pattern. In Beijing, the factors of socioeconomic status and ethnicity remain effective but move to less prominent roles, and the family status factor is almost absent in Beijing. Census data and corresponding spatial data (e.g., TIGER files in the U.S.) are conveniently available for almost any cities in developed countries, and implementing social area analysis in these cities is fairly easy. However, reliable data sources are often a large obstacle for social area studies in cities in developing countries, and future studies can certainly benefit from data of better quality, i.e., data with more socioeconomic, demographic, and housing variables and in smaller geographic units.

APPENDIX 7A: DISCRIMINANT FUNCTION ANALYSIS

Certain categorical things bear some characteristics, each of which can be measured in a quantitative way. The goal of discriminant function analysis (DFA) is to find a linear function of the characteristic variables and use the function to classify future observations into the above known categories. DFA is different from cluster analysis, in which categories are unknown. For example, we know that females and males bear different bone structures. Now some remnants of bodies are found, and we need to identify the gender of the bodies by the DFA.

Here we use a two-category example to illustrate the concept. Say we have two types of objects, A and B, measured by p characteristic variables. The first type has m observations and the second type has n observations. In other words, the observed data are

$$X_{ijA} \ (i = 1, 2, \ldots, m; j = 1, 2, \ldots, p)$$

$$X_{ijB} \ (i = 1, 2, \ldots, n; j = 1, 2, \ldots, p)$$

The objective is to find a discriminant function R such that

$$R = \sum_{k=1}^{p} c_k X_k - R_0 \qquad (A7.1)$$

where $c_k \ (k = 1, 2, \ldots, p)$ and R_0 are constants.

After substituting all m observations X_{ijA} into R, we have m values of $R(A)$. Similarly, we have n values of $R(B)$. The $R(A)$'s have a statistical distribution, and so do the $R(B)$'s. The goal is to find a function R, such that the distributions of $R(A)$ and $R(B)$ are most remote to each other. This goal is met by two conditions, such as:

1. The mean difference $Q = \overline{R(A)} - \overline{R(B)}$ is maximized.
2. The variances $F = S_A^2 + S_B^2$ are minimized (i.e., with narrow bands of distribution curves).

That is equivalent to maximizing $V = Q/F$ by selecting coefficients c_k. Once the c_k's are obtained, we simply use the pooled average of estimated means of $R(A)$ and $R(B)$ to represent R_0:

$$R_0 = [m\overline{R(A)} + n\overline{R(B)}] / (m + n) \qquad (A7.2)$$

For any given sample, we can calculate its R value and compare to R_0. If it is greater than R_0, it is category A; otherwise, it is category B.

DFA is implemented in PROC DISCRIM, or other procedures such as PROC STEPDISC or PROC CANDISC in SAS.

APPENDIX 7B: SAMPLE SAS PROGRAM FOR FACTOR AND CLUSTER ANALYSES

```
/* FA_Clust.SAS runs Factor Analysis & Cluster Analysis
     for social area analysis in Beijing                    */
/* By Fahui Wang on 2-4-05                                   */

/* read the attribute data */
proc import
datafile="c:\gis_quant_book\projects\bj\bjattr.csv"
     out=bj1 dbms=dlm replace;
     delimiter=', ';
     getnames=yes;
proc means;

/* Run factor analysis */
proc factor out=fscore(replace=yes)
     nfact=4 rotate=varimax; /* 4 factors used */
     var x1-x14;
/* export factor score data */
proc export data=fscore dbms=csv
     outfile="c:\gis_quant_book\projects\bj\factscore.csv";

/* Run cluster analysis */
/* Factor scores are first weighted by relative importance
     i.e., variance portions accounted for (based on FA) */
data clust; set fscore;
     factor1 = 0.3516*factor1;
     factor2 = 0.1542*factor2;
```

```
    factor3 = 0.1057*factor3;
    factor4 = 0.0922*factor4;
proc cluster method=ward outtree=tree;
    id ref_id; var factor1-factor4; /*plot dendrogram */
proc tree out=bjclus ncl=9; /*cut the tree at 9 clusters*/
    id ref_id;
/* export the cluster analysis result */
proc export data=bjclus dbms=csv
    outfile="c:\gis_quant_book\projects\bj\cluster9.csv";
run;
```

NOTES

1. Data standardization involves the process of converting a series of data x to a new series x' with mean equal to 0 and standard deviation equal to 1: $x' = (x - \bar{x})/\sigma$.
2. The choice of factor number does not affect the result of PCA. One may start with an arbitrary number (say, 3 or 5) of factors, and results remain the same.

Part III

---

*Advanced Quantitative Methods
and Applications*

8 Geographic Approaches to Analysis of Rare Events in Small Population and Application in Examining Homicide Patterns

When rates are used as estimates for an underlying risk of a rare event (e.g., cancer, AIDS, homicide), those with a small base population have high variance and are thus less reliable. The spatial smoothing techniques, such as the floating catchment area method and the empirical Bayesian smoothing method, as discussed in Chapter 2, can be used to mitigate the problem. This chapter begins with a survey of various approaches to the problem of analyzing rare events in a small population in Section 8.1. Two geographic approaches, namely, the ISD method and the spatial-order method, are fairly easy to implement and are introduced in Section 8.2. The spatial clustering method based on the scale-space theory requires some programming and is discussed in Section 8.3. In Section 8.4, the case study of analyzing homicide patterns in Chicago is presented to illustrate the scale-space melting method implemented in Visual Basic. The section also provides a brief review of the substantive issues: job access and crime patterns. The chapter is concluded in Section 8.5 with a brief summary.

8.1 THE ISSUE OF ANALYZING RARE EVENTS IN A SMALL POPULATION

Researchers in criminology and health studies and others are often confronted with the task of analyzing rare events in a small population and have long sought solutions to the problem.

For criminologists, the study of homicide rates across geographic units and for demographically specific groups often entails analysis of aggregate homicide rates in small populations. Several nongeographic strategies have been attempted by criminologists to mitigate the problem. For example, Morenoff and Sampson (1997) used homicide counts instead of per capita rates or simply deleted outliers or unreliable estimates in areas with a small population. Some used larger units of analysis (e.g., states, metropolitan areas, or large cities) or aggregated over more years to generate stable homicide rates. Land et al. (1996) and Osgood (2000) used

Poisson-based regressions to better capture the nonnormal error distribution pattern in regression analysis of homicide rates in small populations (see Appendix 8).[1]

On the other side, many researchers in health-related fields are well trained in geography and have used several spatial analytical or geographic methods to address the issue. Geographic approaches aim at constructing larger geographic areas, based on which more stable rate estimates may be obtained. The purpose of constructing larger geographic areas is similar to that of aggregating over a longer period of time: to achieve a greater degree of stability in homicide rates across areas. The technique has much common ground with the long tradition of regional classification (*regionalization*) in geography (Cliff et al., 1975). For instance, Black et al. (1996) developed the *ISD method* (after the Information and Statistics Division of the Health Service in Scotland, where it was devised) to group a large number of census enumeration districts (EDs) in the U.K. into larger analysis units of approximately equal population size. Lam and Liu (1996) used the *spatial-order method* to generate a national rural sampling frame for HIV/AIDS research, in which some rural counties with insufficient HIV cases were merged to form larger sample areas. Both approaches emphasize spatial proximity, but neither considers within-area homogeneity of attribute. Haining et al. (1994) attempted to consolidate many EDs in the Sheffield Health Authority Metropolitan District in the U.K. to a manageable number of regions for health service delivery (hereafter referred to as the Sheffield method). The Sheffield method started by merging adjacent EDs sharing similar deprivation index scores (i.e., complying with *within-area attribute homogeneity*), and then used several subjective rules and local knowledge to adjust the regions for spatial compactness (i.e., accounting for *spatial proximity*). The method attempted to balance two criteria (attribute homogeneity and spatial proximity), a major challenge in regionalization analysis. In other words, only contiguous EDs can be clustered together, and these EDs must have similar attributes.

The ISD method and the spatial-order method will be discussed in Section 8.2 in detail. The Sheffield method relies on subjective criteria and involves a substantial amount of manual work that requires one's knowledge of the study area. Section 8.3 will introduce a new spatial clustering method based on the scale-space theory. The method melts adjacent polygons of similar attributes into clusters like the Sheffield method, but is an automated process based on objective criteria. Constructing geographic areas enables the analysis to be conducted at multiple geographic levels, and thus permits the test of the *modifiable areal unit problem* (MAUP).

Table 8.1 summarizes all approaches to the problem of analysis of rates of rare events in a small population.

8.2 THE ISD AND THE SPATIAL-ORDER METHODS

The ISD method is illustrated in Figure 8.1 (based on Black et al., 1996, with modifications). A starting polygon (e.g., the southernmost one) is selected first, and its nearest and contiguous polygon is added. If the total population is equal to or more than the threshold population, the two polygons form an analysis area. Otherwise, the next nearest polygon (contiguous to either of the previous selected polygons) is added. The process continues until the total population of selected

TABLE 8.1
Approaches to Analysis of Rates of Rare Events in a Small Population

	Approach	Examples	Comments
1	Use homicide counts instead of per capita rates	Morenoff and Sampson (1997)	Not applicable for most studies that are interested in the offense or victimization rate relative to population size
2	Delete samples of small populations	Harrell and Gouvis (1994); Morenoff and Sampson (1997)	Deleted observations may contain valuable information
3	Aggregate over more years or to a high geographic level	Messner et al. (1999); most studies surveyed by Land et al. (1990)	Impossible to analyze variations within the time period or within the large areal unit
4	Poisson-based regressions	Osgood (2000); Osgood and Chambers (2000)	Effective remedy for OLS regressions; not applicable to nonregression studies
5	Construct geographic areas with large enough populations	Haining et al. (1994); Black et al. (1996); Sampson et al. (1997)	Generate reliable rates for statistical reports, mapping, regression analysis, and others

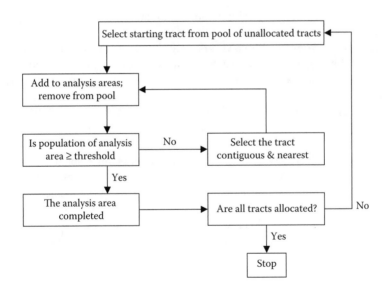

FIGURE 8.1 The ISD method.

polygons reaches the threshold value and a new analysis area is formed. The whole procedure is repeated until all polygons are allocated to new analysis areas. One may use ArcGIS to generate a matrix of distances between polygons and another matrix of polygon adjacency, and then write a simple computer program to implement the method outside of GIS (e.g., Wang and O'Brien, 2005). The method is primitive and does not account for spatial compactness. Some analysis areas

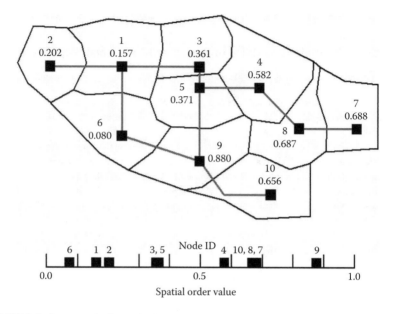

FIGURE 8.2 An example for assigning spatial-order values to polygons.

generated by the method may exhibit odd shapes, and some (particularly those near the boundaries) may require manual adjustment.

The spatial-order method follows a rationale similar to that of the ISD method. It uses space-filling curves to determine the nearness or spatial order of polygons. Space-filling curves traverse space in a continuous and recursive manner to visit all polygons, and assign a spatial order (from 0 to 1) to each polygon based on its relative positions in a two-dimensional space. The procedure, currently available in ArcInfo Workstation, is SPATIALORDER, based on one of the algorithms developed by Bartholdi and Platzman (1988). In general, polygons that are close together have similar spatial-order values and polygons that are far apart have dissimilar spatial-order values. See Figure 8.2 for an example. The method provides a first-cut measure of closeness. The SPATIALORDER command is available in the ArcPlot module through the ArcInfo Workstation command interface. Once the spatial-order value of each polygon is determined, the COLLOCATE command in ArcInfo follows by assigning nearby polygons one group number and accounting for the capacity of each group formed by polygons. Finally, polygons are dissolved based on the group numbers.

8.3 THE SCALE-SPACE CLUSTERING METHOD

The ISD and the spatial-order method only consider spatial proximity, but not within-area attribute homogeneity. The spatial clustering method based on the scale-space theory accounts for both criteria. Development of the scale-space theory has benefited from the advancement of computer image processing technologies, and most of its applications are in analysis of remote sensing data. Here we use the method for addressing the issue of analyzing rare events in small populations.

As we know, objects in the world appear in different ways depending upon the scale of observation. In the case of an image, the size of scale ranges from a single pixel to a whole image. There is no right scale for an object, as any real-world object may be viewed at multiple scales. The operation of systematically simplifying an image at a finer scale and representing it at coarser levels of scale is termed *scale-space smoothing*. A major reason for scale-space smoothing is to suppress and remove unnecessary and disturbing details (Lindeberg, 1994, p. 10). There are various scale-space clustering algorithms (e.g., Wong, 1993; Wong and Posner, 1993). In essence, an image is composed of many pixels with different brightness. As the scale increases, smaller pixels are melted to form larger pixels. The melting process is guided by some objectives, such as entropy maximization (i.e., minimizing loss of information). Applying the scale-space clustering method in a socioeconomic context requires simplification of the algorithm.

The procedures below are based on Wang (2005). The idea is that major features of an image can be captured by its brightest pixels (represented as local maxima). By merging surrounding pixels (up to local minima) to the local maxima, the image is simplified with fewer pixels while the structure is preserved. Five steps implement the concept:

1. *Draw a link between each polygon and its most similar adjacent polygon*: A polygon i has t attributes $(x_{i1}, \ldots, x_{it})$, and its adjacent polygons j $(j = 1, 2, \ldots, m)$ have attributes $(x_{j1}, \ldots, x_{jt})$. Attributes x_{it} and x_{jt} are standardized. Polygon i is linked to only polygon k among its adjacent polygons j based on the rook contiguity (sharing a boundary, not only a vertex) if $D_{ik} = \min_j^m \{ \sum_t (x_{it} - x_{jt})^2 \}$, i.e., the minimum distance criterion.[2] As a result, a link is established between each polygon and one of its adjacent polygons with the most similar attributes.

2. *Determining the link's direction*: The direction of the link between polygons i and k is determined by their attribute values, represented by an aggregate score (Q). In the case study in Section 8.4., Q is the average of three factor scores weighted by their corresponding eigenvalues (representing proportions of variance captured by the factors). Higher scores of any of the three factors indicate more socioeconomic disadvantages. The direction is defined such as $i \rightarrow k$ or $L_{ik} = 1$ if $Q_i < Q_k$; otherwise, $i \leftarrow k$ or $L_{ik} = 0$. Therefore, the directional link always points toward a higher aggregate score. For instance, in Figure 8.3, the arrow points to polygon 1 for the link between 1 and 2 because $Q_2 < Q_1$.

3. *Identifying local minima and maxima*: A local minimum (maximum) is a polygon with all directional links pointing toward other polygons (itself), i.e., with the lowest (highest) Q among surrounding polygons.

4. *Grouping around local maxima*: Beginning with a local minimum, search outward following link directions until a local maximum is reached. All polygons between the local minimum and maximum are grouped into one cluster. If other local minima are also linked to the same local maximum, all polygons along the routes are also grouped into the same cluster. This step is repeated until all polygons are grouped.

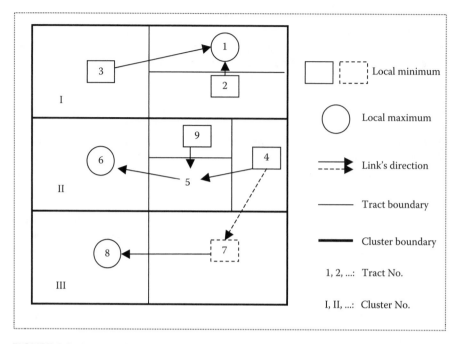

FIGURE 8.3 An example of clustering based on the scale-space theory.

5. *Continuing the next-round clustering*: Steps 1 to 4 yield the result of the first round of clustering, and each cluster can be represented by the averaged attributes of composed polygons (weighted by each polygon's population). The result is fed back to step 1, which begins another round of grouping. The process may be repeated until all units are grouped into one cluster.

Now we use a simple example shown in Figure 8.3 to explain the process. In step 1, polygon 1 is linked to both 2 and 3. Polygons 1 and 3 are linked because 3 is the polygon most similar to 1 between polygon 1's adjacent polygons 2 and 3, but the link between polygons 1 and 2 is established because 1 is the polygon most similar to 2 among polygon 2's adjacent polygons 1, 3, 9, and 4. Similarly, polygon 4 is linked to both 5 and 7, but 5 is the polygon most similar to 4 among polygon 4's adjacent polygons 2, 9, 5, and 7, and 4 is the polygon most similar to 7 among polygon 7's adjacent polygons 4, 5, and 8. Step 2 computes the values of Q for all polygons. In step 3, polygons 2, 3, 4, and 9 are all initially identified as local minima, as all the links are pointed outward there; polygons 1, 6, and 8 are all local maxima, as all the links are pointed inward there. In step 4, both polygons 2 and 3 point to 1, and they are grouped into cluster I; polygons 4 and 9 point to 5 and then to 6, and they are all grouped into cluster II. By doing so, any local maximum (the brightest pixel) serves as the center of a cluster, and surrounding polygons (with less brightness) are melted into the cluster. The cluster stops until it reaches local minima (with the least brightness). The process is repeated until all polygons are grouped.

Note that in Figure 8.3, polygon 4 points to two polygons 5 and 7, but it follows the link to 5 instead of the link to 7 to begin the melting process, as polygon 5 is the most similar one among polygon 4's four adjacent polygons (the link between 4 and 7 is established because 4 is polygon 7's most similar adjacent polygon). Once polygon 4 is melted to cluster II, the link between 4 and 7 becomes redundant and is indicated as a broken link in dashed line, and polygon 7 becomes a new local minimum (indicated in a dashed box). Polygons 7 and 8 are thus grouped together to form cluster III. Also refer to Figure 8.6 for a sample area illustrating the melting process.

The spatial clustering method based on the scale-space theory is implemented in the program file `Scalespace.dll`, developed in Visual Basic.[3] The file is attached in the CD, and its usage is illustrated in the next section. The process may be repeated to generate multiple levels of clustering.

8.4   CASE STUDY 8: EXAMINING THE RELATIONSHIP BETWEEN JOB ACCESS AND HOMICIDE PATTERNS IN CHICAGO AT MULTIPLE GEOGRAPHIC LEVELS BASED ON THE SCALE-SPACE MELTING METHOD

Most crime theories suggest, or at least imply, an inverse relationship between legal and illegal employment. The *strain theory* (e.g., Agnew, 1985) argues that crime results from the inability to achieve desired goals, such as monetary success, through conventional means like legitimate employment. The *control theory* (e.g., Hirschi, 1969) suggests that individuals unemployed or with less desirable employment have less to lose by engaging in crime. The *rational choice* (e.g., Cornish and Clarke, 1986) and *economic* (e.g., Becker, 1968) *theories* argue that people make rational choices to engage in a legal or illegal activity by assessing the cost, benefit, and risk associated with it. Research along this line has focused on the relationship between unemployment and crime rates (e.g., Chiricos, 1987). According to the economic theories, job market probably affects economic crimes (e.g., burglary) more than violent crimes, including homicide (Chiricos, 1987). Support for the relationship between job access and homicide can be found in the *social stress theory*. According to the theory, "high stress can indicate the lack of access to basic economic resources and is thought to be a precipitator of ... homicide risk" (Rose and McClain, 1990, pp. 47–48). Social stressors include any psychological, social, and economic factors that form "an unfavorable perception of the social environment and its dynamics," particularly unemployment and poverty, which are explicitly linked to social problems, including crime (Brown, 1980).

Most literature on the relation between job market and crime has focuses on the link between unemployment and crime using large areas such as the whole nation, states, or metropolitan areas (Levitt, 2001). There may be more variation *within* such units than *between* them. Recent advancements have been made by analyzing the relationship between local job market and crime (e.g., Bellair and Roscigno, 2000). Wang and Minor (2002) argued that not every job is an economic opportunity for all, and only an accessible job is meaningful. They proposed that *job accessibility*, reflecting one's ability to overcome spatial and other barriers to employment, was a better measure of local job market condition. Their study in Cleveland suggested a reverse relationship between job accessibility and crime, and stronger (negative) relationships with economic crimes (including auto theft, burglary, and robbery) than violent crimes (including aggravated assault, homicide, and rape). Wang (2005) further extended the work to focus on the relationship between job access and homicide patterns with refined methodology, based on which this case study is developed. The study focused on homicides for two reasons. First, homicide is considered the most accurately reported crime rate for interunit comparison (Land et al., 1990, p. 923). Second, homicide is rare, and analysis of homicide in small populations makes a good example to illustrate the methodological issues emphasized by this chapter.

This case study uses OLS regressions to examine the possible association between job access and homicide rates in Chicago while controlling for socioeconomic covariates. Case study 9C in Section 9.6 will use spatial regressions to account spatial autocorrelation.

The following datasets are provided in the CD for this project:

1. A polygon coverage `citytrt` contains 845 census tracts (excluding one polygon without any tract code or residents) in the city of Chicago (excluding the O'Hare tract because of its unique land use and noncontiguity with other tracts)
2. A text file `cityattr.txt` contains tract IDs and 10 corresponding socioeconomic attribute values based on the 1990 Census.
3. A program file `Scalespace.dll` implements the scale-space clustering tool.

In the attribute table of coverage `citytrt`, the item `cntybna` is each tract's unique ID, the item `popu` is population in 1990, the item `JA` is job accessibility measured by the methods discussed in Chapter 4 (a higher `JA` value corresponds to better job accessibility), and the item `CT89_91` is total homicide counts for a 3-year period around 1990 (i.e., 1989 to 1991). Homicide data for the study area are extracted from the 1965 to 1995 Chicago homicide dataset compiled by Block et al. (1998), available through the National Archive of Criminal Justice Data (NACJD) at www.icpsr.umich.edu/NACJD/home.html. Homicide counts over a period of 3 years are used to help reduce measurement errors and stabilize rates. In addition, for convenience it also contains the result from the factor analysis (implemented in step 0 below): `factor1`, `factor2`, and `factor3` are scores of three factors that have captured most of the information contained in the socioeconomic attribute file `cityattr.txt`. Note that the job market for defining job accessibility is based on a much wider area (six mostly urbanized counties: Cook, DuPage, Kane, Lake, McHenry, and Will) than the city of Chicago.

Data for defining the 10 socioeconomic variables and population are based on the STF3A files from the 1990 Census and are measured in percentage. In the text file `cityattr.txt`, the first column is tract IDs (i.e., identical to the item `cntybna` in the GIS layer `citytrt`) and the 10 variables are in the following order:

TABLE 8.2
Rotated Factor Patterns of Socioeconomic Variables in Chicago 1990

	Factor 1	Factor 2	Factor 3
Public assistance	0.93120	0.17595	−0.01289
Female-headed households	0.89166	0.15172	0.16524
Black	0.87403	−0.23226	−0.15131
Poverty	0.84072	0.30861	0.24573
Unemployment	0.77234	0.18643	−0.06327
Non-high school diploma	0.40379	0.81162	−0.11539
Crowdedness	0.25111	0.83486	−0.12716
Latinos	−0.51488	0.78821	0.19036
New residents	−0.21224	−0.02194	0.91275
Renter occupied	0.45399	0.20098	0.77222

1. Families below the poverty line (labeled "poverty" in Table 8.2)
2. Families receiving public assistance ("public assistance")
3. Female-headed households with children under 18 ("female-headed households")
4. "Unemployment"
5. Residents who moved in the last 5 years ("new residents")
6. Renter-occupied homes ("renter occupied")
7. Residents without high school diplomas ("no high school diploma")
8. Households with an average of more than 1 person per room ("crowdedness")
9. Black residents ("black")
10. Latino residents ("Latinos")

0. Optional: *Factor analysis on socioeconomic covariates*: Use SAS or other statistical software to conduct factor analysis based on the 10 socioeconomic covariates contained in `cityattr.txt`. Save the result (factor scores and the tract IDs) in a text file and attach it to the GIS layer. This step provides another practice opportunity for principal components and factor analysis, discussed in Chapter 7. Refer to Appendix 7B for a sample SAS program containing a factor analysis procedure. It is optional, as the result (factor scores) is already provided in the polygon coverage `citytrt`.

The principal components analysis result shows that three components (factors) have eigenvalues greater than 1 and are thus retained. These three factors capture 83% of the total variance of the original 10 variables. Table 8.2 shows the rotated factor patterns. Factor 1 (accounting for 56.6% variance among three factors) is labeled "concentrated disadvantage" and captures five variables (public assistance, female-headed households, black, poverty, and unemployment). Factor 2 (accounting for 26.6% variance among three factors) is labeled "concentrated Latino immigration" and captures three variables (residents with no high school diplomas, households with more than one person per room, and Latinos). Factor 3 (accounting for 16.7% variance among three factors) is labeled "residential instability" and captures two variables (residential instability and renter-occupied homes). The three factors are used as control variables (socioeconomic covariates) in the regression analysis of job access and homicide rate. The higher the value of each factor is, the more disadvantageous a tract is in terms of socioeconomic characteristics.

1. *Creating the shapefile with valid census tracts*: Open the coverage `citytrt` in ArcMap > Use Select by Attributes to select polygons with popu > 0 (845 tracts selected) > Export to a shapefile `citytract`.

2. *Computing homicide rates*: Because of rarity of the incidence, homicide rates are usually measured as homicides per 100,000 residents. Open the shapefile `citytract` in ArcMap and open its attribute table > Add a field `homirate` to the table, and calculate it as `homirate = CT89_91 *100000/popu`. The rate is measured as per 100,000 residents.

   In regression analysis, the logarithmic transformation of homicide rates (instead of the raw homicide rate) is often used to measure the dependent variable (see Land et al., 1990, p. 937), and 1 is added to the rates to avoid taking the logarithm of zero.[4] Add another field, `Lhomirat`, to the attribute table of shapefile `citytract` and calculate it as `Lhomirat = log(homirate+1)`.

3. *Mapping tracts with small population*: Figure 8.4 shows that 74 census tracts have a population of fewer than 500, and 28 tracts fewer than 100. Check the raw homicide rate, `homirate`, in these small-population tracts, and note that some tracts have very high rates. This highlights the problem of unstable rates in small populations.

4. *Regression analysis at the census tract level*: Use Microsoft Excel or SAS to run an OLS regression at the census tract level: the dependent variable is `Lhomirat` and the explanatory variables are `JA`, `factor1`, `factor2`, and `factor3`. Refer to Section 6.5.1 if necessary. The result is shown in Table 8.3.

5. *Installing the scale-space clustering tool*: In ArcMap, choose Tools > Customize > click the Command tab > choose "Add from file," browse to the `ScaleSpace.dll` file saved under your project directory, and open it > still with the Command tab clicked in the same dialog window, find and click Scale-Space Tool under Categories to install it.

6. *Using the clustering tool to obtain first-round clusters*: Click the 🆂🆂 button from ArcMap to access the "scale-space cluster" tool and activate the dialog window. Define the choices in the dialog as shown in Figure 8.5. The input shapefile is `citytract`. Use arrows to move variables `factor1`, `factor2`, and `factor3` to the column of "selected fields," which are used as criteria measuring the attribute similarity among tracts. Input their corresponding weights: 0.566, 0.266, and 0.167 (based on the percentage of variance captured by each factor). Use the variable POPU as the weight field to compute weighted averages of attributes in

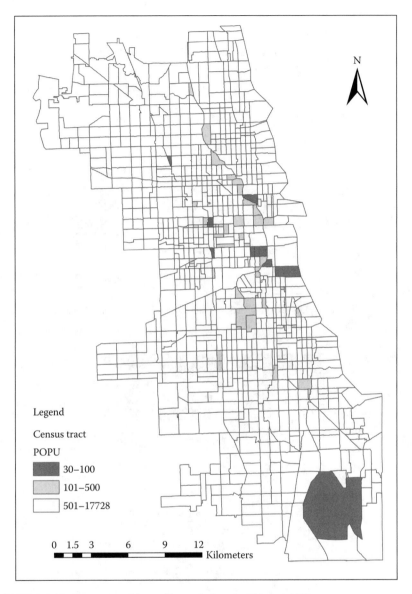

FIGURE 8.4 Census tracts with small populations in Chicago 1990.

the clusters to be formed. The field for the cluster membership in the input shapefile may be named `Clus1` (or others). Define the output directory and the shapefile name (e.g., `Cluster1` by default). One may also check the two boxes for showing and saving intermediate results, and name the shapefile identifying the local minima and local maxima and the shapefile for link directions and types. Finally, click OK to execute the analysis.

TABLE 8.3
OLS Regression Results from Analysis of Homicide
in Chicago 1990

	Census Tracts	First-Round Clusters
No. of observations	845	316
Intercept	6.1324	6.929
	$(10.87)^{***}$	$(8.14)^{***}$
Factor 1	1.2200	1.001
	$(15.43)^{***}$	$(8.97)^{***}$
Factor 2	0.4989	0.535
	$(7.41)^{***}$	$(5.82)^{***}$
Factor 3	−0.1230	−0.283
	(-1.84)	$(-2.93)^{**}$
Job Accessibility (JA)	−2.9143	−3.230
	$(-5.41)^{***}$	$(-3.97)^{***}$
R^2	0.317	0.441

Note: t values in parentheses; ***, significant at 0.001; **, significant at 0.01; *, significant at 0.05.

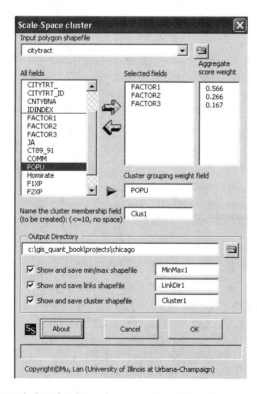

FIGURE 8.5 Dialog window for the scale-space clustering tool.

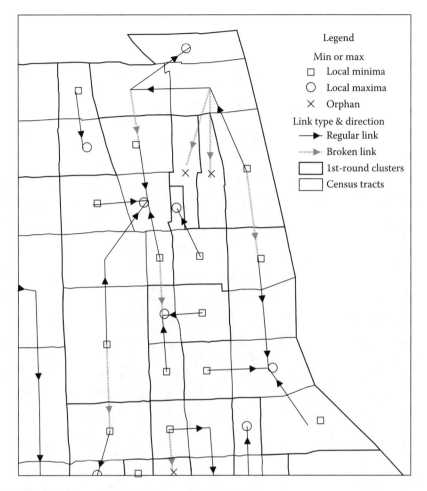

FIGURE 8.6 A sample area for illustrating the clustering process.

Figure 8.6 is the northeast corner of the study area showing the clustering process and result. If no links are pointed from or toward a tract (often as a result of broken links), it is an "orphan" and forms a cluster itself. The clustering result is saved in the shapefile `Cluster`. Additional fields are also created in the attribute table of shapefile `citytract` to save some intermediate results in the clustering process, as well as the clustering result. The attribute table of shapefile `Cluster1` contains the weighted averages of attribute variables `factor1`, `factor2`, and `factor3`, as well as the weight field `POPU`. Figure 8.7 shows the result of this first-round clustering. One may conduct further grouping based on the shapefile `Cluster1`.

7. *Aggregating data to the first-round clusters*: Both the independent and dependent variables (`JA`, `factor1`, `factor2`, `factor3`, and `homirate`) need to aggregate to the cluster level (identified by the field

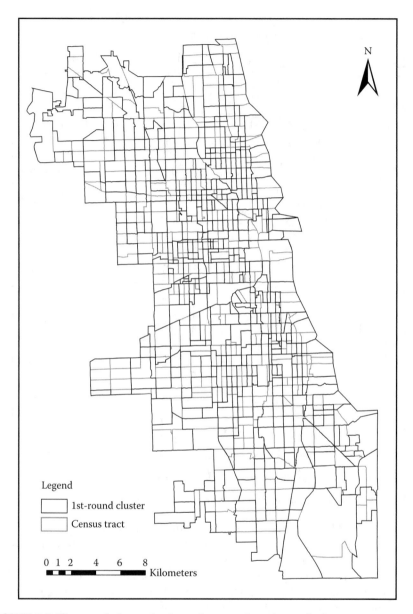

FIGURE 8.7 First-round clusters by the scale-space clustering method.

Clus1) by calculating the weighted averages using the population (popu) as weights.[5] Variables (e.g., factor1, factor2, factor3) in the attribute table of shapefile Cluster1 are already the weighted averages. This step shows how the computation is implemented in ArcMap based on the attribute table of shapefile citytract. Taking factor1 as an example, this is achieved by three steps[6]: (a) calculating a field (say, F1XP) as factor1 multiplied by popu, (b) computing the total population (say, `sum_popu`) and summing up the new field `F1XP` (say, `sum_F1XP`) within each cluster, and (c) dividing `sum_F1XP` by `sum_popu` to obtain the weighted value for `factor1`. In detail, it is implemented as follows:

a. Add new fields `F1XP`, `F2XP`, `F3XP`, `JAXP`, and `HMXP` to the attribute table of shapefile `citytract` and calculate each of them as:

`F1XP=factor1*popu,`

`F2XP=factor2*popu,`

`F3XP=factor3*popu,`

`JAXP=JA*popu,`

`HMXP=homirate*popu;`

b. Sum up these new fields (`F1XP`, `F2XP`, `F3XP`, `JAXP`, and `HMXP`) and the field `popu` by clusters (i.e., the field `Clus1`), and name the output file `sum_clus1.dbf` containing the cluster IDS (`Clus1`), number of tracts within each cluster (`count`), and the summed-up fields (`Sum_F1XP`, `Sum_F2XP`, `Sum_F3XP`, `Sum_JAXP`, `Sum_HMXP`, `Sum_popu`).

c. Add new fields `factor1`, `factor2`, `factor3`, `JA`, and `homirate` to the file `sum_clus1.dbf` and calculate each of them as:

`factor1=Sum_F1XP/Sum_popu,`

`factor2=Sum_F2XP/Sum_popu,`

`factor3=Sum_F3XP/Sum_popu,`

`JA=Sum_JA/Sum_popu,`

`homirate=Sum_HMXP/Sum_popu;`

Finally, add a field `Lhomirat` to `sum_clus1.dbf` and calculate it as `Lhomirat = log(homirate+1)`.

8. *Regression analysis based on the first-round clusters*: Run the OLS regression in Excel or SAS using `sum_clus1.dbf`. The regression result is also presented in Table 8.3.

The OLS regression results based on both the census tracts and first-round clusters show that job accessibility is negatively related to homicide rates in Chicago. Case study 9C in Section 9.6 will further examine the issue while controlling for spatial autocorrelation.

8.5 SUMMARY

In geographic areas with few events (e.g., cancer, AIDS, homicide), rate estimates are often unreliable because of random error associated with small numbers. Researchers have proposed various approaches to mitigate the problem. Applications are particularly rich in criminology and health studies. Among various methods, geographic approaches seek to construct larger geographic areas so that more stable rates may be obtained. The ISD method and the spatial-order method are fairly primitive and do not consider whether areas grouped together are homogenous in attributes. The spatial clustering method based on the scale-space theory accounts for attribute homogeneity while grouping adjacent geographic areas together. It is inevitable that aggregation to larger geographic areas results in the loss of some of the original detail. The scale-space melting process is guided by some objectives, such as entropy maximization (i.e., minimizing loss of information). The method treats a study area composed of many polygons as a picture of pixels. If the attributes in each pixel may be summed up as a single index, this index can be regarded as a measurement of brightness capturing the structure of a picture in black and white. By grouping the pixels together around the brightest ones, fewer and larger pixels are used to capture the basic structure of the original picture at a finer resolution.

A test version of this method is implemented in Visual Basic and incorporated in the ArcGIS environment. The method is applied to examining homicide patterns in Chicago and analyzing whether they are related to job access. The study shows that poorer job access indeed is associated with higher homicide rates while controlling for socioeconomic covariates.

APPENDIX 8: THE POISSON-BASED REGRESSION ANALYSIS

This appendix is based on Osgood (2000). Assuming the timing of the events is random and independent, the *Poisson distribution* characterizes the probability of observing any discrete number $(0, 1, 2, ...)$ of events for an underlying mean count. When the mean count is low (e.g., in a small population), the Poisson distribution is skewed toward low counts. In other words, only these low counts have meaningful probabilities of occurrence. When the mean count is high, the Poisson distribution approaches the normal distribution and a wide range of counts have meaningful probabilities of occurrence.

The *basic Poisson regression model* is

$$\ln(\lambda_i) = \beta_0 + \beta_1 x_1 + \beta_2 x_2 + ... + \beta_k x_k \tag{A8.1}$$

where λ_i is the mean (expected) number of events for case i, x's are explanatory variables, and β's are regression coefficients. Note that the left-hand side in Equation A8.1 is the logarithmic transformation of the dependent variable. The probability of an observed outcome y_i follows the Poisson distribution, given the mean count λ_i, such as

$$\Pr(Y_i = y_i) = \frac{e^{-\lambda_i} \lambda_i^{y_i}}{y_i!} \tag{A8.2}$$

Equation A8.2 indicates that the expected distribution of crime counts depends on the fitted mean count λ_i.

In many studies, it is the rates, not the counts, of events that are of most interest to analysts. Denoting the population size for case i as n_i, the corresponding rate is λ_i / n_i. The regression model for rates is written as

$$\ln(\lambda_i / n_i) = \beta_0 + \beta_1 x_1 + \beta_2 x_2 + ... + \beta_k x_k$$

i.e.,

$$\ln(\lambda_i) = \ln(n_i) + \beta_0 + \beta_1 x_1 + \beta_2 x_2 + ... + \beta_k x_k \tag{A8.3}$$

Equation A8.3 adds the population size n_i (with a fixed coefficient of 1) to the basic Poisson regression model (Equation A8.1) and transforms the model of analyzing counts to a regression model of analyzing rates. The model is a Poisson-based regression that is standardized for the size of base population, and solutions can be found in many statistical packages (e.g., LIMDEP).

Note that the variance of the Poisson distribution is the mean count λ, and thus its standard deviation is $SD_\lambda = \sqrt{\lambda}$. The mean count of events equals the underlying per capita rate r multiplied by the population size n, i.e., $\lambda = rn$. When a variable is divided by a constant, its standard deviation is also divided by the constant. Therefore, the standard deviation of rate r is

$$SD_r = SD_\lambda / n = \sqrt{\lambda} / n = \sqrt{rn} / n = \sqrt{r} / \sqrt{n} \tag{A8.4}$$

Equation A8.4 shows that the standard deviation of per capita rate r is inversely related to the population size n, i.e., the problem of heterogeneity of error variance discussed in Section 8.1. The Poisson-based regression explicitly addresses the issue by acknowledging the greater precision of rates in larger populations.

NOTES

1. Aggregate crime rates from small populations violate two assumptions of ordinary least squares (OLS) regressions, i.e., homogeneity of error variance (because errors of prediction are larger for crime rates in smaller populations) and normal error distribution (because more crime rates of zero are observed as populations decrease).
2. Depending on the applications and the variables used, criteria defining attribute similarity can be different. For example, in the study of regional partitioning of Jingsu Province in China, Luo et al. (2002) computed the correlation coefficients between a county and its adjacent counties and drew a link between the two with the highest coefficient. Their goal was to group areas of a similar socioeconomic structure, i.e., grouping counties at lower development levels with central cities at higher development levels, to form economic regions. As discussed in Section 7.2, there are also different measures for distance.
3. The scale-space cluster tool was developed by Dr. Lan Mu at the Department of Geography, University of Illinois–Urbana-Champaign.

4. The choice of adding 1 (instead of 0.2, 0.5, or others) is arbitrary and may bias the coefficient estimates. However, different additive constants have minimal consequence for significance testing, as standard errors grow proportionally with the coefficients and thus leave the t values unchanged (Osgood, 2000, p. 36). In addition, adding 1 ensures that $\log(r + 1) = 0$ for $r = 0$ (zero homicide).

5. In an updated version of program file **Scalespace.dll**, to be released soon, all variables in the clusters are computed directly by the scale-space cluster tool.

6. In formula, the weighted average is $\bar{x}_w = (\sum w_i x_i) / \sum w_i$.

9 Spatial Cluster Analysis, Spatial Regression, and Applications in Toponymical, Cancer, and Homicide Studies

Spatial cluster analysis detects unusual concentrations or nonrandomness of events in space and time. Nonrandomness of events indicates the existence of *spatial autocorrelation*, and thus necessitates the usage of *spatial regression* in regression analysis of those events. Since the issues were raised several decades ago, applications of spatial cluster analysis and spatial regression were initially limited because of their requirements of intensive computation. Recent advancements in software development, including availability of many free packages, have stimulated greater interests and wide applications. This chapter discusses spatial cluster analysis and spatial regression, and introduces related spatial analysis packages that implement some of the methods.

Two application fields utilize spatial cluster analysis extensively. In crime studies, it is often referred to as hot-spot analysis. Concentrations of criminal activities or hot spots in certain areas may be caused by (1) particular activities, such as drug trading (e.g., Weisburd and Green, 1995); (2) specific land uses, such as skid row areas and bars; or (3) interaction between activities and land uses, such as thefts at bus stops and transit stations (e.g., Block and Block, 1995). Identifying hot spots is useful for police and crime prevention units to target their efforts on limited areas. Health-related research is another field with wide usage of spatial cluster analysis. Does the disease exhibit any spatial clustering pattern? What areas experience a high or low prevalence of disease? Elevated disease rates in some areas may arise simply by chance alone or may be of no public health significance. The pattern generally warrants study only when it is statistically significant (Jacquez, 1998). Spatial cluster analysis is an essential and effective first step in any exploratory investigation. If the spatial cluster patterns of a disease do exist, case-control, retrospective cohort, and other observational studies can follow up.

Rigorous statistical procedures for cluster analysis may be divided into point-based and area-based methods. Point-based methods require exact locations of individual occurrences, whereas area-based methods use aggregated disease rates in regions. Data availability dictates which methods are used. The common belief that point-based methods are better than area-based methods is not well grounded

(Oden et al., 1996). In this chapter, Section 9.1 discusses point-based spatial cluster analysis, followed by a case study of Tai place-names (or *toponymical study*) in southern China using the software SaTScan in Section 9.2. Section 9.3 covers area-based spatial cluster analysis, followed by a case study of cancer patterns in Illinois in Section 9.4. Area-based spatial cluster analysis is implemented by some spatial statistics now available in ArcGIS. Other software, such as CrimeStat (Levine, 2002), provides similar functions. In addition, Section 9.5 introduces spatial regression, and Section 9.6 uses the package GeoDa to illustrate some of the methods in a case study of homicide patterns in Chicago. The chapter is concluded by a brief summary in Section 9.7. Other than ArcGIS, both SaTScan and GeoDa are free software for researchers. There are a wide range of methods for spatial cluster analysis and regression, and this chapter only introduces some exemplary methods, i.e., those most widely used and implemented in the aforementioned packages.

9.1 POINT-BASED SPATIAL CLUSTER ANALYSIS

The methods for point-based spatial cluster analysis can be grouped into two categories: tests for global clustering and tests for local clusters.

9.1.1 POINT-BASED TESTS FOR GLOBAL CLUSTERING

Tests for *global clustering* are used to investigate whether there is clustering throughout the study region. The test by Whittemore et al. (1987) computes the average distance between all cases and the average distance between all individuals (including both cases and controls). *Cases* represent individuals with the disease (or the events in general) being studied, and *controls* represent individuals without the disease (or the nonevents in general). If the former is lower than the latter, it indicates clustering. The method is useful if there are abundant cases in the central area of the study area, but not good if there is a prevalence of cases in peripheral areas (Kulldorff, 1998, p. 53). The method by Cuzick and Edwards (1990) examines the k nearest neighbors to each case and tests whether there are more cases (not controls) than what would be expected under the null hypothesis of a purely random configuration. Other tests for global clustering include Diggle and Chetwynd (1991), Grimson and Rose (1991), and others.

9.1.2 POINT-BASED TESTS FOR LOCAL CLUSTERS

For most applications, it is also important to identify cluster locations or *local clusters*. Even when a global clustering test does not reveal the presence of overall clustering in a study region, there may be some places exhibiting local clusters.

The geographical analysis machine (GAM) developed by Openshaw et al. (1987) first generates grid points in a study region, then draws circles of various radii around each grid point, and finally searches for circles containing a significantly high prevalence of cases. One shortcoming of the GAM method is that it tends to generate a high percentage of false positive circles (Fotheringham and Zhan, 1996). Since many significant circles overlap and contain the same cluster of cases, the Poisson tests that determine each circle's significance are not independent, and thus lead to the problem of multiple testing.

The test by Besag and Newell (1991) only searches for clusters around cases. Say k is the minimum number of cases needed to constitute a cluster. The method identifies the areas that contain the $k - 1$ nearest cases (excluding the centroid case), then analyzes whether the total number of cases in these areas[1] is large relative to the total risk population. Common values for k are between 3 and 6 and may be chosen based on sensitivity analysis using different k values. As in the GAM, clusters identified by Besag and Newell's test often appear as overlapping circles. But the method is less likely to identify false positive circles than the GAM, and is also less computationally intensive (Cromley and McLafferty, 2002, p. 153). Other point-based spatial cluster analysis methods not reviewed here include Rushton and Lolonis (1996) and others.

The following discusses the *spatial scan statistic* by Kulldorff (1997), implemented in SaTScan. SaTScan is a free software program developed by Kulldorff and Information Management Services, available at http://www.satscan.org. Its main usage is to evaluate reported spatial or space-time disease clusters and to see if they are statistically significant.

Like the GAM, the spatial scan statistic uses a circular scan window to search the entire study region, but takes into account the problem of multiple testing. The radius of the window varies continuously in size from 0 to 50% of the total population at risk. For each circle, the method computes the likelihood that the risk of disease is higher inside the window than outside the window. The spatial scan statistic uses either a Poisson-based model or a Bernoulli model to assess statistical significance. When the risk (base) population is available as aggregated area data, the Poisson-based model is used, and it requires case and population counts by areal units and the geographic coordinates of the points. When binary event data for case-control studies are available, the Bernoulli model is used, and it requires the geographic coordinates of all individuals. The cases are coded as ones and controls as zeros.

For instance, under the Bernoulli model, the likelihood function for a specific window z is

$$L(z,p,q) = p^n (1-p)^{m-n} q^{N-n} (1-q)^{(M-m)-(N-n)} \tag{9.1}$$

where N is the total number of cases in the study region, n is the number of cases in the window, M is the total number of controls in the study region, m is the number of controls in the window, $p = n / m$ (probability of being a case within the window), and $q = (N - n) / (M - m)$ (probability of being a case outside the window).

The likelihood function is maximized over all windows, and the "most likely" cluster is one that is least likely to have occurred by chance. The likelihood ratio for the window is reported and constitutes the *maximum likelihood ratio test* statistic. Its distribution under the null hypothesis and its corresponding p value are determined by a Monte Carlo simulation approach. The method also detects secondary clusters with the highest likelihood function for a particular window that do not overlap with the most likely cluster or other secondary clusters.

9.2 CASE STUDY 9A: SPATIAL CLUSTER ANALYSIS OF TAI PLACE-NAMES IN SOUTHERN CHINA

This project extends the toponymical study of Tai place-names in southern China, introduced in Sections 3.2 and 3.4, which focus on mapping the spatial patterns based on spatial smoothing and interpolation techniques. Mapping is merely descriptive and cannot identify whether the concentrations of Tai place-names in some areas are random. The answer relies on rigorous statistical analysis, in this case, point-based spatial cluster analysis. The software SaTScan (the current version is 5.1) is used to implement the study.

The project uses the same datasets as in case studies 3A and 3B: mainly, the point coverage qztai with the item TAI identifying whether a place-name is Tai (= 1) or non-Tai (= 0). In addition, the shapefile qzcnty is provided for mapping the background.

1. *Preparing data in ArcGIS for SaTScan*: Implementing the Bernoulli model for point-based spatial cluster analysis in SaTScan requires three data files: a case file (containing location ID and number of cases in each location), a control file (containing location ID and number of controls in each location), and a coordinates file (containing location ID and Cartesian coordinates or latitude and longitude). The three files can be read by SaTScan through its Import Wizard.

   In the attribute table of qztai, the item TAI already defines the case number (= 1) for each location, and thus the case file. For defining the control file, open the attribute table of qztai in ArcGIS, add a new field NONTAI, and calculate it as NONTAI = 1-TAI. For defining the coordinates file, use ArcToolbox > Coverage Tools > Data Management > Tables > Add XY Coordinates to add X-COORD and Y-COORD. Export the attribute table to a dBase file qztai.dbf.

2. *Executing spatial cluster analysis in SaTScan*: Activate SaTScan and choose Create New Session. A New Session dialog window is shown in Figure 9.1.

   Under the first tab, Input, use the Import Wizard to define the case file: clicking ▤ next to Case File > choose qztai.dbf as the input file > in the SaTScan Input Wizard dialog, choose qztai-id under Source File Variable for Location ID, and similarly TAI for Number of Cases. Define the Control File and the Coordinates File similarly.

   Under the second tab, Analysis, click Purely Spatial under Type of Analysis, Bernoulli under Probability Model, and High Rates under "Scan for Areas with."

   Under the third tab, Output, input Taicluster as the Results File and check all four boxes under dBase.

   Finally, choose Execute Ctl+E under the main menu Session to run the program. Results are saved in various dBase files sharing the file name Taicluster, where the field CLUSTER identifies whether a place is included in a cluster (= 1 for the primary cluster, = 2 for the secondary cluster, = <null> for those not included in a cluster).

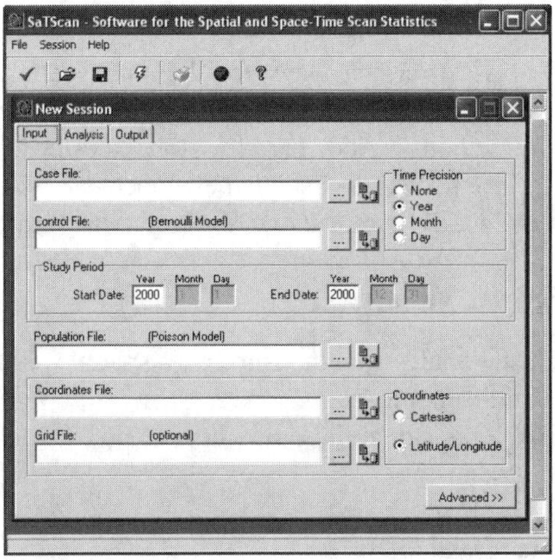

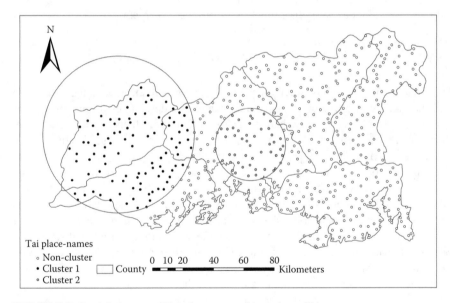

FIGURE 9.1 SaTScan dialog for point-based spatial cluster analysis.

FIGURE 9.2 Spatial clusters of Tai place-names in southern China.

3. *Mapping spatial cluster analysis results*: In ArcGIS, join the dBase file `Taicluster.gis.dbf` to the attribute table of `qztai` using the common key (`LOC_ID` in `Taicluster.gis.dbf` and `qztai-id` in `qztai`). Figure 9.2 uses different symbols to highlight the places that are included in the primary and secondary clusters. The two circles are drawn by hand to show the approximate extents of clusters.

   The spatial cluster analysis confirms that the major concentration of Tai place-names is in the west of Qinzhou, and a minor concentration is in the middle.

9.3 AREA-BASED SPATIAL CLUSTER ANALYSIS

This section first discusses various ways for defining spatial weights, and then introduces two types of statistics available in ArcGIS 9.0. Similarly, area-based spatial cluster analysis methods include tests for global clustering and corresponding tests for local clusters. The former are usually developed earlier than the latter. Other area-based methods include Rogerson's (1999) *R statistic*[2] and others.

9.3.1 DEFINING SPATIAL WEIGHTS

Area-based spatial cluster analysis methods utilize a spatial weights matrix to define spatial relationships of observations.

Defining spatial weights can be based on distance (d):

1. Inverse distance ($1/d$)
2. Inverse distance squared ($1/d^2$)
3. Distance band (= 1 within a specified critical distance and = 0 outside of the distance)
4. A continuous weighting function of distance, such as

$$w_{ij} = \exp(-d_{ij}^2 / h^2)$$

where d_{ij} is the distance between areas i and j, and h is referred to as the *bandwidth* (Fotheringham et al., 2000, p. 111). The bandwidth determines the importance of distance; i.e., a larger h corresponds to a larger sphere of influence around each area. Defining spatial weights can also be based on polygon contiguity (see Section 1.4.2), where $w_{ij} = 1$ if area j is adjacent to i and 0 otherwise.

All the above methods of defining spatial weights can be incorporated in the Spatial Statistics tools in ArcGIS. In particular, the spatial weights are defined at the stage of Conceptualization of Spatial Relationships, which provides the options of Inverse Distance, Inverse Distance Squared, Fixed Distance Band, Zone of Indifference, and Get Spatial Weights From File. All methods based on distance use the geometric centroids to represent areas,[3] and distances are defined as either Euclidean or Manhattan distances. The spatial weights file should contain three columns: from feature ID, to feature ID, and weight (defined as travel distance, time, or cost). The file should be defined prior to the analysis.

The current version of ArcGIS does not incorporate spatial weights based on polygon contiguity. GeoDa provides the option of using rook or queen contiguity to define spatial weights and computes corresponding spatial cluster indexes.

9.3.2 AREA-BASED TESTS FOR GLOBAL CLUSTERING

Moran's I statistic (Moran, 1950) is one of the oldest indicators that detects global clustering (Cliff and Ord, 1973). It detects whether nearby areas have similar or dissimilar attributes overall, i.e., positive or negative *spatial autocorrelation*, respectively. Moran's *I* is calculated as

$$I = \frac{N \sum_i \sum_j w_{ij}(x_i - \bar{x})(x_j - \bar{x})}{(\sum_i \sum_j w_{ij}) \sum_i (x_i - \bar{x})^2} \tag{9.2}$$

where N is the total number of areas, w_{ij} are the spatial weights, x_i and x_j are the attribute values for areas i and j, respectively, and $\bar{x}$ is the mean of the attribute values.

It is helpful to interpret Moran's I as the correlation coefficient between a variable and its *spatial lag*. The spatial lag for variable x is the average value of x in neighboring areas j defined as

$$x_{i,-1} = \sum_j w_{ij} x_j / \sum_j w_{ij} \tag{9.3}$$

Therefore, Moran's I varies between -1 and 1. A value near 1 indicates that similar attributes are clustered (either high values near high values or low values near low values), and a value near -1 indicates that dissimilar attributes are clustered (either high values near low values or low values near high values). If a Moran's I is close to 0, it indicates a random pattern or absence of spatial autocorrelation.

Similar to Moran's I, *Geary's C* (Geary, 1954) detects global clustering. Unlike Moran's I using the cross-product of the deviations from the mean, Geary's C uses the deviations in intensities of each observation with one another. It is defined as

$$C = \frac{(N-1) \sum_i \sum_j w_{ij}(x_i - x_j)^2}{2(\sum_i \sum_j w_{ij}) \sum_i (x_i - \bar{x})^2} \tag{9.4}$$

The values of Geary's C typically vary between 0 and 2, although 2 is not a strict upper limit, with $C = 1$ indicating that all values are spatially independent from each other. Values between 0 and 1 typically indicate positive spatial autocorrelation, while values between 1 and 2 indicate negative spatial autocorrelation, and thus Geary's C is inversely related to Moran's I. Geary's C is sometimes referred to as *Getis–Ord general G* (as is the case in ArcGIS), in contrast to its local version G_i statistic.

Statistical tests for Moran's I and Geary's C can be obtained by means of randomization.

The newly added Spatial Statistics Toolbox in ArcGIS 9.0 provides the tools to calculate both Moran's I and Geary's C. They are available in ArcToolbox > Spatial Statistics Tools > Analyzing Patterns > Spatial Autocorrelation (Moran's I) or High-Low Clustering (Getis–Ord general G). GeoDa and CrimeStat also have the tools for computing Moran's I and Geary's C.

9.3.3 Area-Based Tests for Local Clusters

Anselin (1995) proposed a local Moran index or local indicator of spatial association (LISA) to capture local pockets of instability or local clusters. The local Moran index for an area i measures the association between a value at i and values of its nearby areas, defined as

$$I_i = \frac{(x_i - \bar{x})}{s_x^2} \sum_j [w_{ij}(x_j - \bar{x})] \tag{9.5}$$

where $s_x^2 = \sum_j (x_j - \bar{x})^2 / n$ is the variance and other notations are the same as in Equation 9.2. Note that the summation over j does not include the area i itself, i.e., $j \neq i$. A positive I_i means either a high value surrounded by high values (high–high) or a low value surrounded by low values (low–low). A negative I_i means either a low value surrounded by high values (low–high) or a high value surrounded by low values (high–low).

Similarly, Getis and Ord (1992) developed the local version of Geary's C or the G_i statistic to identify local clusters with statistically significant high or low attribute values. The G_i statistic is written as

$$G_i^* = \frac{\sum_j (w_{ij} x_j)}{\sum_j x_j} \tag{9.6}$$

where the notations are the same as in Equation 9.5, and similarly, the summations over j do not include the area i itself, i.e., $j \neq i$. The index detects whether high values or low values (but not both) tend to cluster in a study area. A high G_i value indicates that high values tend to be near each other, and a low G_i value indicates that low values tend to be near each other. The G_i statistic can also be used for spatial filtering in regression analysis (Getis and Griffith, 2002), as discussed in Appendix 9.

Statistical tests for the local Moran's and local G_i's significance levels can also be obtained by means of randomization.

In ArcGIS 9.0, the tools are available in ArcToolbox > Spatial Statistics Tools > Mapping Clusters > Cluster and Outlier Analysis (Anselin local Moran's I) for computing the local Moran, or Hot Spot Analysis (Getis–Ord G_i^*) for computing the local G_i. The results can be mapped by using the "Cluster and Outlier Analysis with Rendering" tool and the "Hot Spot Analysis with Rendering" tool in ArcGIS. GeoDa and CrimeStat also have the tools for computing the local Moran, but not local G_i.

In analysis for disease or crime risks, it may be interesting to focus only on local concentrations of high rates or the high–high areas. In some applications, all four types of associations (high–high, low–low, high–low, and low–high) revealed by the LISA values have important implications. For example, Shen (1994, p. 177) used the Moran's I to test two hypotheses on the impact of growth control policies in the San Francisco area. The first is that residents who are not able to settle in communities with growth control policies would find the second-best choice in a nearby area, and consequently, areas of population loss (or very slow growth) would be close to areas of fast population growth. This leads to a negative spatial autocorrelation. The second is related to the so-called NIMBY (not in my backyard) symptom. In this case, growth control communities tend to cluster together; so do the pro-growth communities. This leads to a positive spatial autocorrelation.

9.4 CASE STUDY 9B: SPATIAL CLUSTER ANALYSIS OF CANCER PATTERNS IN ILLINOIS

This case study uses the county-level cancer incidence data in Illinois from the Illinois State Cancer Registry (ISCR), Illinois Department of Public Health, available at http://www.idph.state.il.us/about/epi/cancer.htm. The ISCR data are released annually, and each data set contains data for a 5-year span (e.g., 1986 to 1990, 1987 to 1991, and so on). The 1996 to 2000 dataset is used for this case study (and also in Wang, 2004). For demonstrating methodology, cancer counts and rates are simply aggregated to the county level without adjustment by age, sex, race, and other factors. The study will examine four cancers with the highest incidence rates: breast, lung, colorectal, and prostate cancers. Along with the cancer registry data, the Illinois Department of Public Health also provides the population data for all Illinois counties in each year. Population for each county during the 5-year period of 1996 to 2000 is simply the average over 5 years.

The data are processed and provided in a coverage ilcnty. In addition to items identifying counties, the five items needed for analysis are POPU9600 (average population from 1996 to 2000), COLONC (5-year count of colorectal cancer incidents), LUNGC (5-year count of lung cancer incidents), BREASTC (5-year count of breast cancer incidents), and PROSTC (5-year count of prostate cancer incidents).

1. *Computing and mapping cancer rates*: Open the attribute table of ilcnty in ArcGIS and add fields COLONRAT, LUNGRAT, BREASTRAT, and PROSTRAT. Taking COLONRAT as an example, it is computed as COLONRAT = 100000*COLONC/POPU9600. In other words, the cancer rate is measured as the number of incidents per 100,000. Table 9.1 summarizes the basic statistics for cancer rates at the county level in Illinois from 1996 to 2000. Note that the state rate is obtained by dividing the total cancer incidents by the total population in the whole state, and is different from the mean of cancer rates across counties.[4]

   The following analysis also uses colorectal cancer as an example for illustration. Figure 9.3 shows the colorectal cancer rates in Illinois counties

TABLE 9.1
Cancer Incident Rates (per 100,000) in Illinois Counties, 1986–2000

Cancer Type	State Rate	Mean	Minimum	Maximum	Std. Dev.
Breast — invasive (females)	351.23	384.43	225.59	596.59	66.28
Lung	349.09	446.77	228.73	758.82	119.38
Colorectal	288.30	374.60	205.93	584.13	80.66
Prostate	316.82	369.09	198.74	533.26	83.33

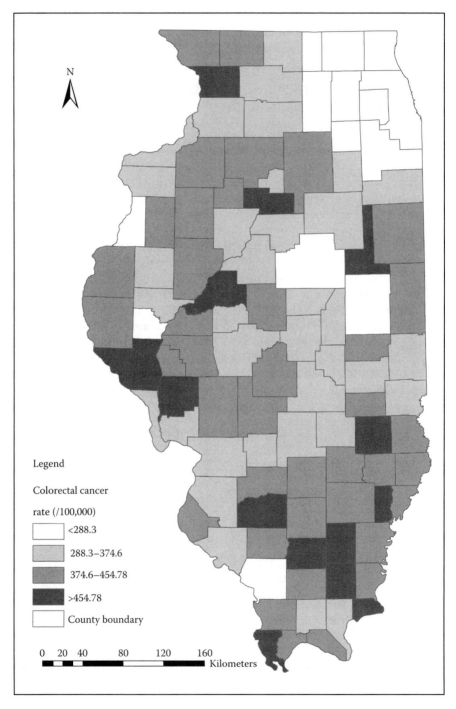

FIGURE 9.3 Colorectal cancer rates in Illinois counties, 1996–2000.

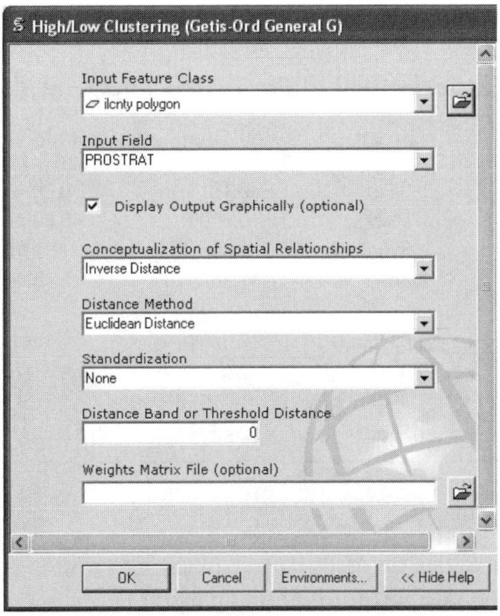

FIGURE 9.4 ArcGIS dialog for computing Getis–Ord general G.

from 1996 to 2000. The first category shows the counties with rates below the state average (288.3), which are mainly concentrated in the Chicago metropolitan area in the northeast corner. The second category shows the counties with rates between the state rate (288.3) and the average rate across counties (374.6). High colorectal cancer rates are observed at the southeast corner, and to a lesser degree in the west.

2. *Computing Getis–Ord general* G *and Moran's* I: In ArcToolbox, choose Spatial Statistics Tools > Analyzing Patterns > High-Low Clustering (Getis–Ord general G) to activate a dialog window shown in Figure 9.4. Choose `ilcnty` (polygon) as the Input Feature Class and `COLONRAT` as the Input Field, and check the option Display Output Graphically (other default choices, such as Inverse Distance for Conceptualization of Spatial Relationships, are okay). The graphic window shows that there is "less than 5% likelihood that this clustered pattern could be the result of random chance." Related statistics are reported in Table 9.2.

Repeat the analysis using the Spatial Autocorrelation (Moran's I) tool. Based on Moran's I, the clustered pattern is even more significant (at the 1% level).

For either the Getis–Ord general G or the Moran's I, the statistical test is a normal z test, such as $z = (Index - Expected)/\sqrt{variance}$. If z is larger than 1.960 (critical value), it is statistically significant at the 0.05 (5%) level, and if z is larger than 2.576 (critical value), it is statistically significant at the 0.01 (1%) level. For instance, for the colorectal cancer rates, the Moran's I is 0.09317, its expected value is -0.0099, and the variance is 0.0001327, and thus $z = (0.09317 - (-0.0099))/\sqrt{0.0001327} = 8.9489$ (i.e., larger than 2.576), indicating the significance above 1%.

TABLE 9.2
Global Clustering Indexes for County-Level Cancer Incident Rates

Index Statistics		Breast	Lung	Colorectal	Prostate
Moran's I	Value	0.0426	0.1211	0.0932	0.0696
	Expected	−0.0099	−0.0099	−0.0099	−0.0099
	Variance	1.3234E-4	1.330E-4	1.3270E-4	1.3384E-4
	Z score	4.5619***	11.3630***	8.9489***	6.8706***
General G	Value	2.0320E-6	2.0508E-6	2.0411E-6	2.0402E-6
	Expected	2.0186E-6	2.0186E-6	2.0186E-6	2.0186E-6
	Variance	7.3044E-17	1.7702E-16	1.1436E-16	1.2590E-17
	Z score	1.5662	2.4209*	2.0993*	1.9257

Note: ***, significant at 0.001; **, significant at 0.01; *, significant at 0.05.

Repeat the analysis on other cancer rates. The results are summarized in Table 9.2. The z values for both the general G and the Moran's I suggest that the spatial clustering pattern is strongest in lung cancer, followed by colorectal, prostate, and breast cancers. The statistical significance is weaker by the general G than by the Moran's I.

3. *Computing local Moran's and local* G_i: In ArcToolbox, use Spatial Statistics Tools > Mapping Clusters > Cluster and Outlier Analysis (Anselin local Moran's I) to activate the dialog. Define the Input Feature Class and the Input Field similar to those in step 2, and name the output layer Colon_Lisa. In the output attribute table, four new fields are added: LMiInvDst is the local Moran's based on inverse distance (for spatial relationship), LMzInvDst the corresponding z value, ExpectedI the expected value, and Variance the variance. The local Moran's can be mapped either directly using the field LMiInvDst or using another tool, "Cluster and Outlier Analysis with Rendering." The index simply reveals the clusters of areas with similar cancer rates (high values) and the clusters of areas with heterogeneous cancer rates (low values). As we are interested in clusters of elevated cancer rates, it is helpful to first exclude the counties with rates below the state rate (288.3), and then highlight those clusters of counties with higher cancer rates. Figure 9.5 shows that the major clusters are at the southeast corner.

   Repeat the analysis using the tool Hot Spot Analysis (Getis–Ord $Gi*$). A new field GiInvDst is created to save the $Gi*$ values in the output layer. A high $Gi*$ value indicates that high cancer rates tend to be near each other (hot spots), and a low G_i value indicates that low cancer rates tend to be near each other (cold spots). Figure 9.6 shows the spatial pattern of colorectal cancer: hot spots in the southeast, cold spots in the northeast, and the areas between.

   The tool Hot Spot Analysis (Getis–Ord $Gi*$) does not generate the z values for the $Gi*$. One needs to use the tool Hot Spot Analysis with Rendering for obtaining the z scores and mapping the results.

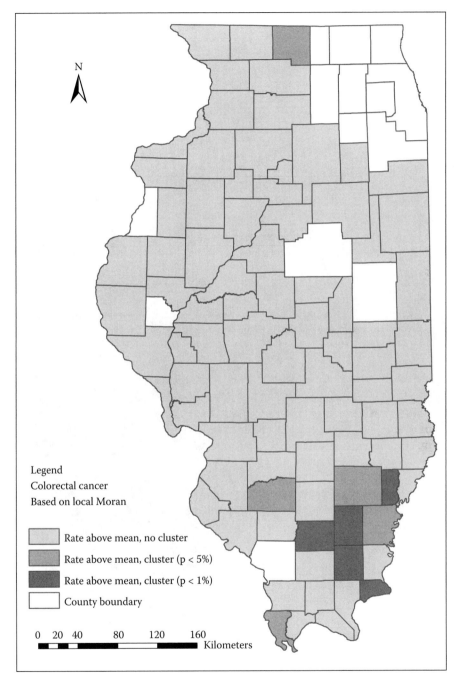

FIGURE 9.5 Colorectal cancer clusters based on local Moran.

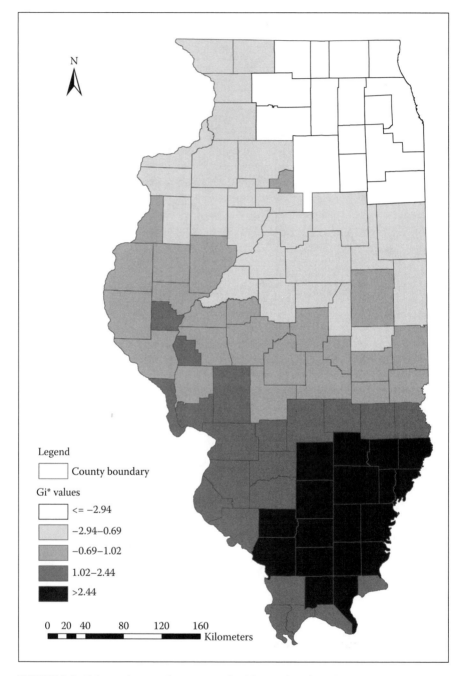

FIGURE 9.6 Colorectal cancer hot spots and cold spots based on *Gi**.

9.5 SPATIAL REGRESSION

The spatial cluster analysis detects *spatial autocorrelation*, in which values of a variable systematically related to geographic location. In the absence of spatial autocorrelation or spatial dependence, the ordinary least squares (OLS) regression model can be used. It is expressed in matrix form:

$$\mathbf{y} = \mathbf{X}\boldsymbol{\beta} + \boldsymbol{\varepsilon} \qquad (9.7)$$

where $\mathbf{y}$ is a vector of n observations of the dependent variable, $\mathbf{X}$ is an $n \times m$ matrix for n observations of m independent variables, $\boldsymbol{\beta}$ is a vector of regression coefficients, and $\boldsymbol{\varepsilon}$ is a vector of random errors or residuals, which are independently distributed about a mean of zero.

When spatial dependence is present, the residuals are no longer independent from each other, and the OLS regression is no longer applicable. This section discusses two commonly used models of *maximum likelihood estimator*. The first is a *spatial lag model* (Baller et al., 2001) or *spatially autoregressive model* (Fotheringham et al., 2000, p. 167). The model includes the mean of the dependent variable in neighboring areas (i.e., *spatial lag*) as an extra explanatory variable. Denoting the weights matrix by $\mathbf{W}$, the spatial lag of $\mathbf{y}$ is written as $\mathbf{Wy}$ as defined in Equation 9.3. The element of $\mathbf{W}$ in the i-th row and j-th column is $w_{ij} / \sum_{j} w_{ij}$. The model is expressed as

$$\mathbf{y} = \rho\mathbf{Wy} + \mathbf{X}\boldsymbol{\beta} + \boldsymbol{\varepsilon} \qquad (9.8)$$

where ρ is the regression coefficient for the spatial lag and other notations are the same as in Equation 9.7.

Rearranging Equation 9.8 yields

$$(\mathbf{I} - \rho\mathbf{W})\mathbf{y} = \mathbf{X}\boldsymbol{\beta} + \boldsymbol{\varepsilon}$$

Assuming the matrix $(\mathbf{I} - \rho\mathbf{W})$ is invertible, we have

$$\mathbf{y} = (\mathbf{I} - \rho\mathbf{W})^{-1}\mathbf{X}\boldsymbol{\beta} + (\mathbf{I} - \rho\mathbf{W})^{-1}\boldsymbol{\varepsilon} \qquad (9.9)$$

This reduced form shows that the value of y_i at each location i is determined not only by x_i at that location (like in the OLS regression model), but also by the x_j at other locations through the spatial multiplier $(\mathbf{I} - \rho\mathbf{W})^{-1}$ (not present in the OLS regression model). The model is also different from the autoregressive model in time series analysis and cannot be calibrated by the SAS procedures for time series modeling, such as AR or AMAR.

The second is a *spatial error model* (Baller et al., 2001) or *spatial moving average model* (Fotheringham et al., 2000, p. 169) or *simultaneous autoregressive* (SAR) *model* (Griffith and Amrhein, 1997, p. 276). Instead of treating the dependent variable as autoregressive, the model considers the error term as autoregressive. The model is expressed as

$$y = X\beta + u \qquad (9.10)$$

where u is related to its spatial lag, such as

$$u = \lambda W u + \varepsilon \qquad (9.11)$$

where λ is a spatial autoregressive coefficient and the second error term ε is independent.

Solving Equation 9.11 for u and substituting into Equation 9.10 yields the reduced form

$$y = X\beta + (I - \lambda W)^{-1}\varepsilon \qquad (9.12)$$

This shows that the value of y_i at each location i is affected by the stochastic errors ε_j at all other locations through the spatial multiplier $(I - \lambda W)^{-1}$.

Estimation of either the spatial lag model in Equation 9.9 or the spatial error model in Equation 9.12 is implemented by the maximum likelihood (ML) method (Anselin and Bera, 1998). The case study in the next section illustrates how the spatial lag and the spatial error models are implemented in GeoDa using the algorithms developed by Smirnov and Anselin (2001). Anselin (1988) discusses the statistics to decide which model to use. The statistical diagnosis rarely suggests that one model is preferred over the other (Griffith and Amrhein, 1997, p. 277).

9.6 CASE STUDY 9C: SPATIAL REGRESSION ANALYSIS OF HOMICIDE PATTERNS IN CHICAGO

This case study continues the analysis of homicide patterns in Chicago introduced earlier. Case study 8 in Section 8.4 used OLS regression, and this project uses spatial regression to control for spatial autocorrelation.

In addition to the polygon coverage `citytrt` used in case study 8, a polygon coverage `citycom` with 77 community areas for the same study area (excluding the O'Hare Airport area) is provided on the CD for this project.

See Section 8.4 for a detailed description of attributes contained in the coverage `citytrt`. In addition, the item `comm` in the attribute table of `citytrt` identifies which community area a tract belongs to. Each of the 77 community areas in the coverage `citycom` is made of multiple whole census tracts. We will analyze the relationship between job access and homicide rate in Chicago at two geographic levels (census tracts and community areas). One may also repeat the spatial regression analysis based on the new analysis units generated by the scale-space clustering analysis in case study 8. The analysis unit increases in area size from census tracts to the first-round clustered areas, to the second-round clustered areas, and to community

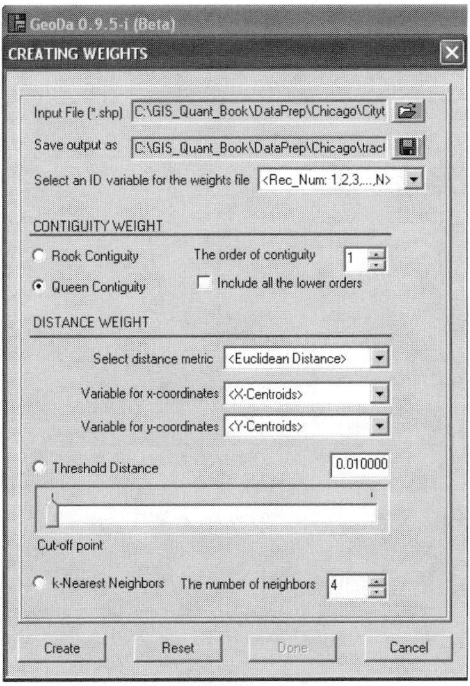

FIGURE 9.7 GeoDa dialog for defining spatial weights.

areas, providing a complete spectrum of areal units in the analysis of homicide patterns in Chicago.

9.6.1 PART 1: SPATIAL REGRESSION ANALYSIS AT THE CENSUS TRACT LEVEL BY GEODA

1. *Preparing for spatial regression*: If one starts the project without completing case study 8, follow the instructions to finish steps 1 and 2 in Section 8.4 to create a shapefile `citytract` with valid census tracts and compute logarithms of homicide rates (field name `Lhomirat`).

2. *Defining spatial weights in GeoDa*: Start GeoDa. Choose Tools > Weights > Create to activate the dialog window for defining the spatial weights (Figure 9.7). In the dialog, select `citytract.shp` as the Input shapefile, enter `tract` as the Output file, choose Queen Contiguity (as an example) to define the spatial weights, and finally click Create to execute. A spatial weights file `tract.GAL` is created.

3. *Running OLS regression in GeoDa*: In GeoDa, choose `Methods` > `Regress`. In the Get DBF File dialog window, select `citytract.dbf` as the Input file name. In the next dialog window, Regression Title & Output, enter "OLS Regression for Census Tracts" as Report Title and

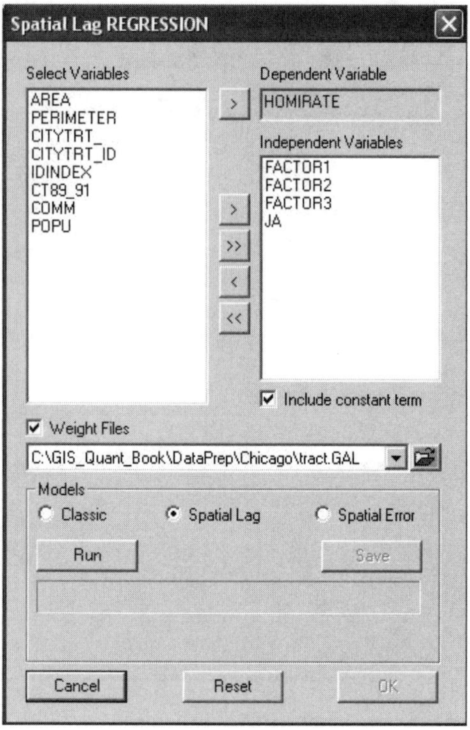

FIGURE 9.8 GeoDa dialog for spatial regression.

"Trt_OLS" as Output file name, and click OK to invoke the model-building dialog, as shown in Figure 9.8. In the new dialog window, (1) use the >, », <, and « buttons to move the variable `homirate` from the dropdown list to the Dependent Variable box, and move the variables `factor1`, `factor2`, `factor3`, and `JA` from the dropdown list to the Independent Variable box; (2) under Models, choose the radio button next to Classic; and (3) click Run to execute it. The result is shown and saved in the file `Trt_OLS.OLS`. See Table 9.3. It is identical to the regression result obtained in SAS and presented in Table 8.3.

4. *Running the spatial lag regression model in GeoDa*: The process for the spatial lag model is essentially the same as for the OLS regression in step 3. Enter a different regression title and output file name. In the model-building dialog window, the differences are (1) under Weight Files, click on the file-open symbol to select `tract.GAL` as the spatial weights file, and (2) choose the radio button next to Spatial Lag. The result is also summarized in Table 9.3.

5. *Running the spatial error regression model in GeoDa*: Follow the same process to run the spatial error model (choose the radio button next to Spatial Error). The result is also reported in Table 9.3. Note that both spatial regression models are obtained by the maximum likelihood (ML) estimation.

TABLE 9.3
OLS and Spatial Regressions of Homicide Rates in Chicago
(n = 845 Census Tracts)

Independent Variables	OLS Model	Spatial Lag Model	Spatial Error Model
Intercept	6.1324	4.5338	5.8304
	(10.87)***	(7.52)***	(8.97)***
Factor 1	1.2200	0.9654	1.1777
	(15.43)***	(10.91)***	(12.89)***
Factor 2	0.4989	0.4048	0.4777
	(7.41)***	(6.01)***	(6.01)***
Factor 3	–0.1230	–0.0993	–0.0858
	(–1.84)	(–1.53)	(–1.09)
Job access	–2.9143	–2.2056	–2.6321
	(–5.41)***	(–4.13)***	(–4.26)***
Spatial lag (ρ)		0.2750	
		(5.90)***	
Spatial error (λ)			0.2627
			(4.82)***
Sq. corr.	0.395	0.424	0.415

Note: t values in parentheses; ***, significant at 0.001; **, significant at 0.01; *, significant at 0.05.

9.6.2 PART 2: SPATIAL REGRESSION ANALYSIS AT THE COMMUNITY AREA LEVEL BY GEODA

1. *Creating the shapefile with valid community areas*: Open the coverage `citycom` in ArcMap > Use Select by Attributes to select polygons with `popu > 0` (77 community areas selected) > Export to a shapefile `citycomm`.
2. *Aggregating data to community areas*: Both the dependent variable (`homirate`) and independent variables (`factor1`, `factor2`, `factor3`, JA) need to be aggregated from the census tracts to the corresponding community areas. Refer to step 7 in Section 8.4, if needed. Join the table containing the summarized result (weighted averages) to the shapefile `citycomm` by the common key `comm`.
3. *Defining spatial weights and running regressions*: Follow steps 2 to 5 in Part 1 to define a new spatial weights file `comm.GAL` based on the shapefile `citycomm`, and run the OLS, spatial lag, and spatial error regression models. The regression results are summarized in Table 9.4.

9.6.3 DISCUSSION

Several observations may be made from the regression results presented in Table 9.3 and Table 9.4.

TABLE 9.4
OLS and Spatial Regressions of Homicide Rates in Chicago
(n = 77 Community Areas)

Independent Variables	OLS Model	Spatial Lag Model	Spatial Error Model
Intercept	5.5679	4.2516	5.3882
	(5.63)***	(4.07)***	(5.11)***
Factor 1	1.2415	1.0671	1.2185
	(8.92)***	(7.22)***	(8.44)***
Factor 2	0.4287	0.4095	0.4244
	(3.45)***	(3.54)***	(3.37)***
Factor 3	−0.3641	−0.3055	−0.3657
	(−3.40)**	(−2.95)**	(−3.20)**
Job access	−1.4246	−1.0768	−1.2599
	(−1.48)	(−1.20)	(−1.23)
Spatial lag (ρ)		0.2369	
		(2.45)*	
Spatial error (λ)			0.1647
			(1.01)
Sq. corr.	0.750	0.769	0.755

Note: t values in parentheses; ***, significant at 0.001; **, significant at 0.01; *, significant at 0.05.

1. In the models for census tracts, both t statistics for the spatial lag (ρ) and the spatial error (λ) are very significant, and thus indicate the necessity of using spatial regression models over the OLS regression. In the models for community areas, the t statistic is significant at 0.05 for the spatial lag (ρ), but not so for the spatial error (λ), and thus indicates that spatial autocorrelation is not as strong as in the case of census tracts. That is to say, running the OLS regression using the community areas risks less model-building error.

2. In the models for both the census tracts and community areas, results (signs and significance levels of coefficients of independent variables) from the spatial regressions are similar to those from the OLS regressions.

3. In the models for the census tracts, areas with poorer job accessibility are associated with higher homicide rates, and the relationship is statistically significant. An earlier study in Cleveland, using bivariate regressions, has shown a consistent inverse relationship between job accessibility and various crime rates (Wang and Minor, 2002). Results from this study provide even stronger evidence as covariates are controlled for.

4. The relationship between job accessibility and homicide rates remains negative in the models for the community areas, but no longer significant. One possible explanation is that community areas in Chicago are defined mainly by geographic features (rivers, railroads and freeways, etc.), and are not necessarily made of homogenous census tracts. As a result, variation of variables is smoothed out within community areas and much

information is lost in data aggregation. Therefore, it is evident that the modifiable area unit problem (MAUP) is present as the analysis unit changes from census tracts to community areas.

5. Among the three factors as covariates, both factors 1 and 2 have expected signs (+) and are statistically significant in all models at both the census tract and community area levels. Factor 3 is not statistically significant in the models for census tracts, but significant in the models for community areas, indicating presence of MAUP.

9.7 SUMMARY

Spatial cluster analysis detects nonrandomness of spatial patterns or existence of spatial autocorrelation. In practice, methods for point-based data and for area-based data are distinct. Point-based methods analyze whether events within a radius exhibit a higher level of concentration than a random pattern would suggest. Area-based methods examine whether objects in proximity or adjacency are related (similar or dissimilar) to each other. Applications of spatial cluster analysis are often seen in crime- and health-related studies. In this chapter, case study 9A applies the point-based spatial cluster analysis technique to analyzing Tai place-names in southern China. One reason for choosing this case study is to demonstrate how GIS-based spatial analysis techniques can be used in fields that are less exposed to the methodologies, such as history and linguistics. Case study 9B illustrates how the area-based spatial cluster analysis methods are used to detect cancer cluster patterns in Illinois.

The existence of spatial autocorrelation necessitates the usage of spatial regression in regression analysis. Two typical models for spatial regression are the spatial lag model and the spatial error model. Both need to be estimated by the maximum likelihood (ML) method. Case study 9C uses the spatial regression models to examine whether job access is related to homicide patterns in Chicago.

The current version of ArcGIS provides some newly included spatial statistics for area-based spatial cluster analysis, but does not have any tools for implementing point-based spatial cluster analysis or spatial regression. For the latter, one may use some free software, such as SaTScan and GeoDa, for research purposes. These packages are designed for some specific research tasks and are usually easy to learn and implement, as demonstrated in this chapter.

APPENDIX 9: SPATIAL FILTERING METHODS FOR REGRESSION ANALYSIS

The spatial filtering methods by Getis (1995) and Griffith (2000) take a different approach to account for spatial autocorrelation in regression. The methods separate the spatial effects from the variables' total effects and allow analysts to use conventional regression methods such as OLS to conduct the analysis (Getis and Griffith, 2002). Compared to the maximum likelihood spatial regression, the major advantage of spatial filtering methods is that the results uncover individual spatial and nonspatial component contributions and are easy to interpret. Griffith's (2000) eigenfunction decomposition method involves intensive computation and takes more steps to implement. This appendix discusses Getis's method.

The basic idea in Getis's method is to partition each original variable (spatial autocorrelated) into a filtered nonspatial variable (spatial independent) and a residual spatial variable, and then feed the filtered variables into OLS regression. Based on the G_i statistic in Equation 9.6, the filtered observation x_i^* is defined as

$$x_i^* = \frac{W_i / (n-1)}{G_i} x_i$$

where x_i is the original observation, $W_i = \sum_j w_{ij}$ (averaged spatial weights for $i \neq j$), n is the number of observations, and G_i is the local G_i statistic. Note that the numerator $W_i / (n-1)$ is the expected value for G_i. When there is no autocorrelation, $x_i^* = x_i$. The difference $L_{xi} = x_i - x_i^*$ represents the spatial component of the variable at i.

Feeding the filtered variables (including the dependent and explanatory variables) into an OLS regression yields the spatially filtered regression model, such as

$$y^* = f(x_1^*, x_2^*, ...)$$

where y^* is the filtered dependent variable and x_1^*, x_2^*, and others are the filtered explanatory variables.

The final regression model includes both the filtered nonspatial component and the spatial component of each explanatory variable, such as

$$y = f(x_1^*, L_{x_1}, x_2^*, L_{x_2}, ...)$$

where y is the original dependent variable and L_{x_1}, L_{x_2}, ... are the corresponding spatial components of explanatory variables x_1, x_2, ...

Like the G_i statistic, Getis's spatial filtering method is only applicable to variables with a natural origin and positive values, not those represented by standard normal variates, rates, or percentage change (Getis and Griffith, 2002, p. 132).

NOTES

1. The number may be slightly larger than k since the last (farthest) area among those nearest areas may contain more than one case.
2. The R statistic is simply a spatial version of the well-known Chi-square goodness-of-fit statistic and is easy to code in a computer program (Wang, 2004).
3. If centroids must be within feature boundaries, use the tool Features to Points and choose the option Inside to create centroids before the analysis (see Section 1.4.1).
4. This is caused by the uneven distribution of base population in computing cancer rates. For instance, counties in the Chicago metropolitan area have large population sizes, and thus exert a dominant effect on the state rates (but a much smaller effect on average rates across counties).

10 Linear Programming and Applications in Examining Wasteful Commuting and Allocating Health Care Providers

This chapter introduces *linear programming* (LP), an important optimization technique in socioeconomic analysis and planning. LP seeks to maximize or minimize an objective function subject to a set of constraints. Both the objective and the constraints are expressed in linear functions. It would certainly take more than one chapter to cover all issues in LP, and many graduate programs in geography, planning, or other fields use a whole course or more to teach LP. This chapter discusses the basic concepts of LP and emphasizes how LP problems are solved in SAS and ArcGIS.

Section 10.1 reviews the formulation of LP and the simplex method. The method is applied to examining the issue of wasteful commuting in Section 10.2. Commuting is an important research topic in urban studies for its theoretical linkage to urban structure and land use, as well as its implications in public policy. Themes in the recent literature of commuting have moved beyond the issue of wasteful commuting and cover a diverse set of issues, such as the relation between commuting and urban land use, explanation of intraurban commuting, and implications of commuting patterns in spatial mismatch and job access. However, strong research interests in commuting are, to some degree, attributable to the issue of wasteful commuting raised by Hamilton (1982). A case study in Columbus, OH, is used to illustrate the method of measuring wasteful commuting, and a SAS program is developed to solve the LP in the case study.

Section 10.3 introduces integer linear programming (ILP), in which some of the decision variables in a linear programming problem take on only integer values. Some classic location-allocation problems, such as the *p*-median problem, the location set covering problem (LSCP), and the maximum covering location problem (MCLP), are used to illustrate the formulation of ILP problems. Applications of these location-allocation problems can be widely seen in both private and public sectors. Section 10.4 uses a hypothetical example of allocating health care providers in Cuyahoga County, Ohio, to illustrate the implementation of a location-allocation problem in ArcGIS. The chapter is concluded in Section 10.5 with a brief summary.

10.1   LINEAR PROGRAMMING (LP) AND THE SIMPLEX ALGORITHM

10.1.1   THE LP STANDARD FORM

The *linear programming* (LP) *problem* in the *standard form* can be described as follows: find the maximum of $\sum_{j=1}^{n} c_j x_j$ subject to the constraints $\sum_{j=1}^{n} a_{ij} x_j \le b_i$ for all $i \in \{1,2,...,m\}$ and $x_j \ge 0$ for all $j \in \{1,2,...,n\}$.

The function $\sum_{j=1}^{n} c_j x_j$ is the *objective function*, and a solution x_j ($j \in \{1,2,...,n\}$) is also called the *optimal feasible point*.

In the matrix form, the problem is stated as follows: Let $\mathbf{c} \in \mathbf{R}^n$, $\mathbf{b} \in \mathbf{R}^m$, and $\mathbf{A} \in \mathbf{R}^{m \times n}$. Find the maximum of $\mathbf{c}^T \mathbf{x}$ subject to the constraints $\mathbf{A}\mathbf{x} \le \mathbf{b}$ and $\mathbf{x} \ge 0$.

Since the problem is fully determined by the data $\mathbf{A}, \mathbf{b}, \mathbf{c}$, it is referred to as problem $(\mathbf{A}, \mathbf{b}, \mathbf{c})$.[1]

Other problems not in the standard form can be converted to it by the following transformations (Kincaid and Cheney, 1991, p. 648):

1. Minimizing $\mathbf{c}^T \mathbf{x}$ is equivalent to maximizing $-\mathbf{c}^T \mathbf{x}$.

2. A constraint $\sum_{j=1}^{n} a_{ij} x_j \ge b_i$ is equivalent to $-\sum_{j=1}^{n} a_{ij} x_j \le -b_i$.

3. A constraint $\sum_{j=1}^{n} a_{ij} x_j = b_i$ is equivalent to $\sum_{j=1}^{n} a_{ij} x_j \le b_i$, $-\sum_{j=1}^{n} a_{ij} x_j \le -b_i$.

4. A constraint $\sum_{j=1}^{n} |a_{ij} x_j| \le b_i$ is equivalent to $\sum_{j=1}^{n} a_{ij} x_j \le b_i$, $-\sum_{j=1}^{n} a_{ij} x_j \le b_i$.

5. If a variable x_j can also be negative, it is replaced by the difference of two variables, such as $x_j = u_j - v_j$.

10.1.2   THE SIMPLEX ALGORITHM

The *simplex algorithm* (Dantzig, 1948) is widely used for solving linear programming problems. By skipping the theorems and proofs, we move directly to illustrate the method in an example.

Consider a linear programming problem in the standard form:

Maximize: $z = 4x_1 + 5x_2$

Subject to: $2x_1 + x_2 \le 12$

$-4x_1 + 5x_2 \le 20$

$$x_1 + 3x_2 \leq 15$$

$$x_1 \geq 0, x_2 \geq 0$$

As the problem is not in the standard form, it needs to be converted to the form before the algorithm is applied. The simplex method begins with introducing *slack variables* $\mathbf{u} \geq 0$ so that the constraints $\mathbf{Ax} \leq \mathbf{b}$ can be converted to an equation form $\mathbf{Ax} + \mathbf{u} = \mathbf{b}$.

For the above problem, three slack variables ($x_3 \geq 0, x_4 \geq 0, x_5 \geq 0$) are introduced. The problem is rewritten as

Maximize: $z = 4x_1 + 5x_2 + 0x_3 + 0x_4 + 0x_5$

Subject to: $2x_1 + x_2 + x_3 + 0x_4 + 0x_5 = 12$

$$-4x_1 + 5x_2 + 0x_3 + x_4 + 0x_5 = 20$$

$$x_1 + 3x_2 + 0x_3 + 0x_4 + x_5 = 15$$

$$x_1 \geq 0, x_2 \geq 0, x_3 \geq 0, x_4 \geq 0, x_5 \geq 0$$

The simplex method is often accomplished by exhibiting the data in a *tableau* form, such as

4	5	0	0	0	
2	1	1	0	0	12
−4	5	0	1	0	20
1	3	0	0	1	15

The top row contains coefficients in the objective function $\mathbf{c}^T\mathbf{x}$. The next m rows represent the constraints that are reexpressed as a system of linear equations. We leave the element at the top-right corner blank, as the solution to the problem z_{max} is yet to be determined. The tableau is of a general form

$$\mathbf{c}^T \quad 0$$
$$\mathbf{A} \quad \mathbf{I} \quad \mathbf{b}$$

If the problem has a solution, it is found at a finite stage in the algorithm. If the problem does not have a solution (i.e., an unbounded problem), the fact is discovered in the course of the algorithm. The tableau is modified in successive steps according to certain rules until the solution (or no solution for an unbounded problem) is found.

By assigning 0 to the original variables x_1 and x_2, the initial solution ($x_1 = 0$, $x_2 = 0$, $x_3 = 12$, $x_4 = 20$, $x_5 = 15$) certainly satisfies the constraints of equations. The variables x_j that are zero are designated *nonbasic variables*, and the remaining ones, usually nonzero, are designated *basic variables*. The tableau has n components of nonbasic variables and m components of basic variables, corresponding to the number of original and slack variables, respectively. In the example, x_1 and x_2 are the nonbasic variables ($n = 2$), and x_3, x_4, and x_5 are the basic variables ($m = 3$). In the matrix that defines the constraints, each basic variable occurs in only one row, and the objective function must be expressed only in terms of nonbasic variables.

In each step of the algorithm, we attempt to increase the objective function by converting a nonbasic variable to a basic variable. This is done through Gaussian elimination steps since elementary row operations on the system of equations do not alter the set of solutions. The following summarizes the work on any given tableau:

1. Select the variable x_s whose coefficient in the objective function is the largest positive number, i.e., $c_s = \max\{c_i > 0\}$. This variable becomes the new basic variable.
2. Divide each b_i by the coefficient of the new basic variable in that row, a_{ij}, and among those with $a_{is} > 0$ (for any i), select the minimum b_i / a_{ij} and assign it to the new basic variable, i.e., $x_s = b_k / a_{kj} = \min\{b_i / a_{ij}\}$. If all a_{is} are 0, the problem has no solution.
3. Using pivot element a_{ks}, create zeros in column s with Gaussian elimination steps (i.e., keeping the kth row with the pivot element and subtracting it from other rows).
4. If all coefficients in the objective function (the top row) are ≤ 0, the current x is the solution.

We now apply the procedures to the example.

In step 1, x_2 becomes the new basic variable because 5 (i.e., its coefficient in the objective function) is the largest positive coefficient.

In step 2, a_{22} is identified as the pivot element (highlighted in underscore in the following tableau) because 20/5 is the minimum among $\{12/1, 20/5, 15/3\}$ and $x_2 = 20/5 = 4$.

0.8	1	0	0	0	
2	1	1	0	0	12
−0.8	1	0	0.2	0	4
0.3333	1	0	0	0.3333	5

In step 3, Gaussian eliminations yield a new tableau:

1.6	0	0	−0.2	0	
2.8	0	1	−0.2	0	8
−0.8	1	0	0.2	0	4
1.1333	0	0	−0.2	0.3333	1

According to step 4, the process continues as $c_1 = 1.6 > 0$.

Similarly, x_1 is the new basic variable, a_{13} is the pivot element, $x_1 = 1/1.1333 = 0.8824$, and the resulting new tableau is

0	0	0	0.0515	−0.2941	
0	0	0.3571	0.1050	−0.2941	1.9748
0	1.25	0	0.0735	0.2941	5.8824
1	0	0	−0.1765	0.2941	0.8824

The process continues and generates a new tableau, such as

0	0	−3.4	0	−2.9143	
0	0	3.4	1	−2.8	18.8
0	17	−3.4	0	6.8	61.2
5.6667	0	3.4	0	−1.1333	23.8

By now, all coefficients in the objective function (the top row) are ≤ 0, and the solution is $x_1 = 23.8/5.6667 = 4.2$ and $x_2 = 61.2/17 = 3.6$. The maximum value of the objective function is $z_{max} = 4*4.2 + 5*3.6 = 34.8$.

Many software packages (some free) are available for solving LP problems.[2] We select the LP procedure in SAS (available in the SAS/OR module) to illustrate the implementation of solving LP problems (see Section 10.2.3). SAS, already introduced in previous chapters, is a powerful package and is particularly convenient for coding large matrices. The LP procedure in SAS solves linear programs, integer programs, and mixed-integer problems.

10.2 CASE STUDY 10A: MEASURING WASTEFUL COMMUTING IN COLUMBUS, OHIO

10.2.1 THE ISSUE OF WASTEFUL COMMUTING AND MODEL FORMULATION

The issue of wasteful commuting was first raised by Hamilton (1982). Assuming that residents can freely swap houses, the planning problem is to minimize total commuting given the locations of houses and jobs. Hamilton used the exponential function to capture the distribution patterns of both residents and employment. Since employment is more centralized than population (i.e., the employment density function has a steeper gradient than the population density function), the solution to the problem is that commuters always travel toward the CBD and stop at the nearest employer. See Appendix 10A for details. Hamilton found 87% wasteful commuting in 14 U.S. cities. White (1988) proposed a simple LP model to measure wasteful commuting. White's study yielded very little wasteful commuting, likely attributable to the large area unit she used (Small and Song, 1992). Using a smaller unit, Small and Song (1992) applied White's LP approach to Los Angeles and found about 66% wasteful commuting, less than that in Hamilton's model but still substantial. The following formulation follows White's LP approach.

Given the number of resident workers P_i at i ($i = 1, 2, ..., n$) and the number of jobs E_j at j ($j = 1, 2, ..., m$), the minimum commute is the solution to the following linear programming problem:

Minimize: $$\sum_{i=1}^{n}\sum_{j=1}^{m} c_{ij}x_{ij}$$

Subject to: $$\sum_{j=1}^{m} x_{ij} \leq P_i \text{ for all } (i = 1, 2, ..., n)$$

$$\sum_{i=1}^{n} x_{ij} \leq E_j \text{ for all } (j = 1, 2, ..., m)$$

$$x_{ij} > 0 \text{ for all } (i = 1, 2, ..., n) \text{ and all } (j = 1, 2, ..., m)$$

where c_{ij} is the commute distance (time) from residential location i to job site j, and x_{ij} is the number of commuters on that route.

The objective function is the total amount of commute in the city. The first constraint defines that the total commuters from each residential location to various job locations cannot exceed the number of resident workers there. The second constraint defines that the total commuters from various residential locations to each job site cannot exceed the number of jobs there. In the urbanized areas of most U.S. metropolitan areas, it is most likely that the total number of jobs exceeds the total number of resident workers, i.e., $\sum_{i=1}^{n} P_i \leq \sum_{j=1}^{m} E_j$.

10.2.2 DATA PREPARATION IN ARCGIS

The following datasets are prepared and provided in the CD for the project:

1. Coverage `urbtazpt` has 991 traffic analysis zone (TAZ) centroids for the study area, i.e., the urbanized area in Columbus MSA in 1990.
2. Coverage `road` contains all roads for the study area.

The spatial datasets are processed from the TIGER files. Specifically, the coverage `urbtazpt` is defined by combining all TAZs within the Columbus MSA, including seven counties (Franklin, Union, Delaware, Licking, Fairfield, Pickaway, and Madison), and extracting the urbanized portion in 1990. Wang (2001b) used the same study area in his study of explaining intraurban commute. See Figure 10.1. The coverage `road` is defined similarly by combining all roads in the study area. For maintaining network connectivity, it covers a slightly larger area than the urbanized area. In the attribute table of `urbtazpt`, the item `emp` is the number of jobs in each TAZ and `work` is the number of resident workers. In addition, the item `popu` is the population in each TAZ for reference. These attributes are extracted from the 1990 Census Transportation Planning Package (CTPP) Urban Element

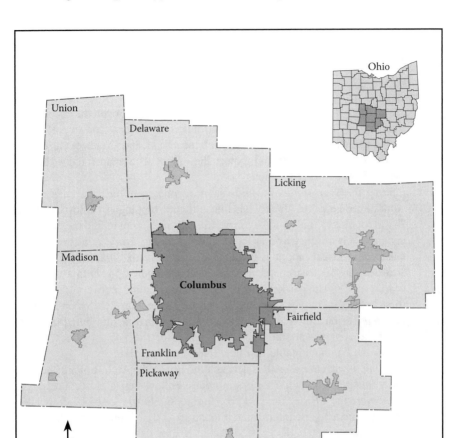

FIGURE 10.1 Columbus MSA and the study area.

datasets for Columbus MSA. Specifically, the information for resident workers and population is based on Part 1 (by place of residence), and the information for jobs is based on Part 2 (by place of work).

The items `emp` and `work` in `urbtazpt` define the numbers of jobs and resident workers in each TAZ, respectively, and thus the variables P_i ($i = 1, 2, ..., n$) and E_j ($j = 1, 2, ..., m$) in the above LP problem. The major task for data preparation is the computation of commute time c_{ij} through the road network. Chapter 2 has discussed the concepts and methods related to distance and travel time measurements, and this case study provides another opportunity to practice the technique. One may skip the steps and go directly to wasteful commuting analysis in SAS, as the resulting files are also provided in the CD for your convenience.

The following explains the procedures for computing commute time in ArcGIS. Please refer to Chapter 2, Sections 2.2 and 2.3, if necessary. This study uses a simple approach for measuring travel time by assuming a uniform speed on the same level of roads. For a more advanced method, see Wang (2003).

1. *Extracting job and resident work locations*: In ArcMap, open the coverage `urbtazpt`, (1) select the TAZ centroids with `emp>0` and export the output to a shapefile `emptaz` (931 job locations), and (2) select the TAZ centroids with `work>0` and export the output to a shapefile `restaz` (812 resident worker locations).

2. *Finding the nearest nodes from resident worker locations*: The road network coverage `road` already has line and node topologies (if not, use the "build topology" tool to do so). First, use the "near" tool to find the nearest node to each resident worker location by choosing `restaz` as the input feature and `road` (node) as the near feature. Note that *unlike* case study 2 in Section 2.3, each resident worker location corresponds to a unique nearest node. This makes the process of computing network distance (time) slightly simpler than that illustrated in Section 2.3. Second, join the *node* attribute table of `road` to `restaz` so that the attribute table of `restaz` contains additional identifications of the nearest nodes, and export the combined table to a new table `resid.dbf`.

3. *Defining the origin nodes*: It is necessary to access the ArcInfo Workstation for defining an INFO file for origin nodes. Follow step 4 in Section 2.3.2 to convert the dBase file `resid.dbf` to INFO file `resid`, drop all items but `road_id`, and rename it to `road-id`.[3] There are 812 unique origin nodes.

4. *Finding the nearest nodes from job locations and defining the destination nodes*: Repeat steps 2 and 3 to define an INFO file `empid` for destination nodes based on the job location layer `emptaz`. Similarly, each job location corresponds to a unique nearest node, and thus there are 931 unique destination nodes.

5. *Defining the impedance for the road network*: Add an item `speed` to the attribute table of `road` (*node*). Similar to the speed limit guidelines in Table 5.1, we assign different speeds to corresponding CFCC codes:

   a. CFCC >= 'A11' and CFCC <= 'A18' speed = 55
   b. CFCC >= 'A21' and CFCC <= 'A28' speed = 45
   c. CFCC >= 'A31' and CFCC <= 'A38' speed = 35
   d. All others speed = 25

   Add another item, `time`, to the table and compute it as `time = (length/speed)*60/1609` in minutes (note the unit for length is meter).

6. *Computing network travel time in ArcPlot*: The following commands in ArcInfo Workstation compute the network travel time and save the result in an INFO file `odtime`:

```
ap                                /* access the ArcPlot modulde
netcover road colsroute           /* set up the route system
centers resid                     /* define the origin nodes
stops empid                       /* define the destination nodes
impedance time                    /* define the impedance item
nodedistance centers stops odtime 600 network ids
q                                 /* exit
```

7. *Attaching air distance segments and summing up travel times*: Join the table resid.dbf (containing the air distance between resident worker locations and their nearest nodes) to odtime, and also join the table empid.dbf (containing the air distance between job locations and their nearest nodes) to odtime. Add a new item comtime, and compute it as odtime: comtime = odtime:network + ((empid.near_dist+resid. near_dist)/25)*60/1609.[4] The INFO file contains $812 \times 931 = 755{,}972$ records.

8. *Exporting data for wasteful commuting analysis in SAS*: Export the attribute table of urbtazpt to a space-delimited text file urbtaz.txt (without headings for variable names) containing values of taz, work and emp. Export the INFO file odtime to a space-delimited text file odtime.txt (without headings for variable names) containing values of resid.taz, empid.taz, and comtime. Both text files will be used for the wasteful commuting analysis in SAS.

An AML program (rdtime.aml) included in the CD implements the whole process, from step 1 to 8.

10.2.3 Measuring Wasteful Commuting in SAS

The definition of an LP problem in SAS takes two formats: a dense and a sparse format. The sparse input format will be illustrated here because of its flexibility of coding large matrices. In the sparse input format, use the COEF, COL, TYPE, and ROW statements to identify variables in the problem dataset, or simply use SAS internal variable names: _COEF_, _COL_, _TYPE_, and _ROW_, respectively.

The following SAS program solves the example illustrated in Section 10.1:

```
Data;
Input _row_ $1-6 _coef_ 8-9 _type_ $11-13 _col_ $15-19;
   Cards;
Object . max .               /* define type for Obj_func */

Object 4 . x1                /* coefficient for 1st variable in Obj_func */

Object 5 . x2                /* coefficient for 2nd variable in Obj_func */

Const1 12 le _RHS_           /* Type & RHS value in 1st constraint */

Const1 2 . x1                /* coefficient for 1st var in 1st constraint */
```

```
Const1  1  .  x2          /* coefficient for 2nd var in 1st constraint */
Const2  20  le  _RHS_     /* Type & RHS value in 2nd constraint */
Const2  -4  .  x1         /* coefficient for 1st var in 2nd constraint */
Const2  5  .  x2          /* coefficient for 2nd var in 2nd constraint */
Const3  15  le  _RHS_     /* Type & RHS value in 3rd constraint */
Const3  1  .  x1          /* coefficient for 1st var in 3rd constraint */
Const3  3  .  x2          /* coefficient for 2nd var in 3rd constraint */
  ;
Proc  lp  sparsedata;     /* run the LP procedure */
Run;
```

The main part in the above SAS program is to code the objective function and constraints in the LP problem. Each takes multiple records defined by four variables:

1. _row_ labels whether it is the objective function or which constraint.
2. _coef_ defines the coefficient of a variable (or a missing value in the first record for coding the objective function type MAX or MIN, or the value on the right-hand side in the first record for coding each constraint).
3. _type_ takes the value MIN or MAX in the first record for coding the objective function, or the value LE, LT, EQ, GE, or GT in the first record for coding each constraint, or a missing value in others.[5]
4. _col_ is the variable name (or a missing value in the first record for coding the objective function, or _RHS_ in the first record for each constraint).

As shown in the above sample program, the first record defines the type for objective function or the right-hand-side value in a constraint. Therefore, the number of records for defining the objective function or each constraint is generally one more than the number of variables.

Coding the wasteful commuting case study in SAS involves more coding complexity, mainly the DO routines. For convenience, the following two data files generated from step 8 in Section 10.2.2 are already provided in the CD:

1. `urbtaz.txt` contains three columns of values (with no headings for variable names): TAZ codes, numbers of resident workers, and numbers of jobs for 991 TAZs.
2. `odtime.txt` contains three columns of values (with no headings for variable names): resident worker TAZ codes, job TAZ codes, and commute times (in minutes) between them.

The `LP.SAS` program in Appendix 10B (also included in the CD) implements the linear programming problem with comments explaining critical steps. After the data of employment, resident workers and commute distances are read into SAS, the program first defines the constraint for employment, then the constraints for resident workers, and finally the objective function. The result is written to an external file `min_com.txt` containing the optimal solution: the origin TAZ code, destination TAZ code, number of commuters, and commute time on each route.

The LP output shows the total minimum commute time of 1,939,404.01 minutes, i.e., an average of 3.90 minutes per resident worker (with a total of 497,588 commuters). The actual mean commute time was 20.40 minutes in the study area in 1990 (Wang, 2001b, p. 173), implying an astonishing 80.89% wasteful commuting. Further examining the optimal commuting patterns in `min_com.txt` reveals that many of the trips have the same origin and destination TAZs, i.e., intrazonal trips. Studies on the 1990 CTPP Part 3 (journey to work) for the study area show that among the 346 TAZs with nonzero intrazonal commuters, the average intrazonal drove-alone time was actually 16.3 minutes. Wang (2003, p. 258) also revealed significant intrazonal commute time in Cleveland (11.3 minutes). This is not captured by the above simple method of network travel time computation, which yields intrazonal commute times under 1 minute.[6]

10.3 INTEGER PROGRAMMING AND LOCATION-ALLOCATION PROBLEMS

10.3.1 GENERAL FORMS AND SOLUTIONS

If some of the decision variables in a linear programming problem take on only integer values, the problem is referred to as an *integer programming* problem.

If all decision variables are integers, it is an *integer linear programming* (ILP) *problem*. Similar to the LP standard form, it is written as

Maximize: $\sum_{j=1}^{n} c_j x_j$

Subject to: $\sum_{j=1}^{n} a_{ij} x_j \le b_i$ for all $i \in \{1,2,...,m\}$

integers $x_j \ge 0$ for all $j \in \{1,2,...,n\}$

If some of the decision variables are integers and others are regular nonnegative numbers, it is a *mixed-integer linear programming* (MILP) *problem*. It is written as

Maximize: $\sum_{j=1}^{n} c_j x_j + \sum_{k=1}^{p} d_k y_k$

Subject to: $\sum_{j=1}^{n} a_{ij} x_j + \sum_{k=1}^{p} g_{ik} y_k \le b_i$ for all $i \in \{1,2,...,m\}$

integers $x_j \ge 0$ for all $j \in \{1,2,...,n\}$

$y_k \ge 0$ for all $k \in \{1,2,...,p\}$

One may think that an easy approach to an ILP or MILP problem is to solve the problem as a regular LP problem and round the solution. In many situations, the rounded solution is not necessarily optimal (Wu and Coppins, 1981, p. 399). This is particularly true when the integer decision variables are restricted to the values 0 or 1, i.e., the *0-1 (binary) programming problem*. The 0-1 programming problem has wide applications in operations research and management science, particularly in location-allocation problems.

Solving the ILP or MILP requires special approaches, such as the *cutting planes method* or the *branch-and-bound method*; the latter is more popular. The following summarizes a general branch-and-bound algorithm.

1. Find a feasible solution f_L as the lower bound on the maximum value of the objective function.
2. Select one of the remaining subsets and separate it into two or more new subsets of solutions.
3. For each subset, compute an upper bound f_U on the maximum value of the objective function over all completions.
4. A subset is eliminated (fathomed) if (1) $f_U < f_L$, or (2) its solution is not feasible, or (3) the best feasible solution in this subset has been found, in which case f_L is replaced with this new value.
5. Stop if there are no remaining subsets. Otherwise, go to step 2.

10.3.2 Location-Allocation Problems

We use three classic location-allocation problems to illustrate the formulation of ILP problems.

The first is the *p-median problem* (ReVelle and Swain, 1970). The objective is to locate a given number of facilities among a set of candidate facility sites so that the total travel distance or time to serve the demands assigned to the facilities is minimized. The *p*-median model formulation is

Minimize: $$Z = \sum \sum a_i d_{ij} x_{ij}$$

Subject to: $x_{ij} \le x_{jj} \ \forall i, j, i \ne j$ (each demand assignment is restricted to what has been located)

$$\sum_{j=1}^{m} x_{ij} = 1 \ \forall i$$ (each demand must be assigned to a facility)

$$\sum_{j=1}^{m} x_{jj} = p \ \forall j$$ (exactly p facilities are located)

$x_{ij} = 0,1 \ \forall i, j$ (a demand area is assigned to only one facility) where i indexes demand areas $(i = 1, 2, \ldots, n)$, j indexes candidate facility sites $(j = 1, 2, \ldots, m)$, p is the number of facilities to be located, a_i is the amount of demand at area i, d_{ij} is the distance or time between demand i and facility j, and x_{ij} is 1 if demand i is assigned to facility j or 0 otherwise.

The second is the *location set covering problem* (LSCP) that minimizes the number of facilities needed to cover all demand (Toregas and ReVelle, 1972). The model formulation is

Minimize: $Z = \displaystyle\sum_{j=1}^{m} x_j$

Subject to: $\displaystyle\sum_{j=1}^{N_i} x_j \geq 1 \ \forall i$ (a demand area must be within the critical distance or time of at least one open facility site)

$x_j = 0,1 \ \forall j$ (a candidate facility is either open or closed)

where N_i is the set of facilities where the distance or time between demand i and facility j is less than the critical distance or time s, i.e., $d_{ij} < s$; x_j is 1 if a facility is open at candidate site j or 0 otherwise; and i, j, m, and n are the same as in the above p-median model formulation.

The third is the *maximum covering location problem* (MCLP) that maximizes the demand covered within a desired distance or time threshold by locating p facilities (Church and ReVelle, 1974). The model formulation is

Minimize: $Z = \displaystyle\sum_{i=1}^{n} a_i y_i$

Subject to: $\displaystyle\sum_{j=1}^{N_i} x_j + y_i \geq 1 \ \forall i$ (a demand area must be within the critical distance or time of at least one open facility site or it is not covered)

$\displaystyle\sum_{j=1}^{m} x_j = p$ (exactly p facilities are located)

$x_j = 0,1 \ \forall j$ (a candidate facility is either open or closed)

$y_i = 0,1 \ \forall i$ (a demand area is either covered or not)

where y_i is 1 if a demand area i is not covered or 0 otherwise, i, j, m, n, and p are the same as in the p-median model formulation, and N_i and x_j are the same as in the above LSCP model formulation.

All of the above three problems may be solved in SAS, but take some programming time to code the problems. It is more convenient to use some built-in models in ArcInfo Workstation (currently not available in ArcGIS desktop) to solve

TABLE 10.1
Location-Allocation Models

Location-Allocation Model	Objective	Constraints	ArcInfo Model
I. p-median problem	Minimize total distance (time)	Locate p facilities; all demands are covered	MINDISTANCE
II. p-median with a maximum distance constraint	Minimize total distance (time)	[Additional to I] Demand must be within a specified distance (time) of its assigned facility	MINDISTANCE (constrained)
III. Location set covering problem (LSCP)	Minimize the number of facilities	All demands are covered	n/a
IV. Maximum covering location problem (MCLP)	Maximize coverage	Locate p facilities; demand is covered if it is within a specified distance (time) of a facility	MAXCOVER
V. MCLP with mandatory closeness constraints	Maximize coverage	[Additional to IV] Demand not covered must be within a second (larger) distance (time) of a facility	MAXCOVER (constrained)

them. The *p-median problem* is solved by the MINDISTANCE model in ArcInfo. The original *p*-median model minimizes total travel effort but does not necessarily limit an individual's travel, which may not be appropriate for some location-allocation problems. One may add a constraint that each demand site must be served by a facility within a critical distance or time. This formulation is known as the *p-median problem with a maximum distance constraint* (Khumawala, 1973; Hillsman and Rushton, 1975). In ArcInfo, it is solved by the MINDISTANCE (constrained) model. Currently ArcInfo does not have a built-in model to solve the *location set covering problem* (LSCP). The *maximum covering location problem* (MCLP) is solved by the MAXCOVER model in ArcInfo. Similarly, one may add an additional constraint to the original MCLP that requires an uncovered demand point within a mandatory closeness constraint (the second and a larger distance threshold). The revised model is known as the *MCLP with mandatory closeness constraints* (Church and ReVelle, 1974) and is solved by the MAXCOVER (constrained) model in ArcInfo. Table 10.1 summarizes the models and corresponding ArcInfo commands. For details, use the ArcInfo online help on the topic Working with Linear Features > Location-Allocation.

The current location-allocation models available in ArcGIS are limited in capacities and do not have the flexibility of adding complex constraints. Many applications call for the formulation of more advanced models that require the use of SAS or other specialized software. See Curtin et al. (2005) for a recent example of applying the location-allocation models to defining optimal police patrol force in Dallas.

10.4 CASE STUDY 10B: ALLOCATING HEALTH CARE PROVIDERS IN CUYAHOGA COUNTY, OHIO

This is a *hypothetical* project developed to illustrate the implementation of location-allocation models in ArcInfo. Say the Public Health Department of Cuyahoga County, Ohio, plans to set up five temporal clinics for administering free flu shots for children (under the age of 5) and seniors (65 years old and over). Sites for the five clinics will be selected from the 21 hospitals in the county. For the purpose of illustration, the objective in this project is to minimize the total travel distance or time for all clients, and thus a *p*-median problem. Other location-allocation models are implemented similarly in ArcInfo.

The following datasets are provided for this project:

1. A polygon coverage `cuyatrt` for all census tracts in Cuyahoga County
2. A shapefile `tgr39035lka` for all roads and streets in the county
3. A comma-separated value file `cuya_hosp.csv` containing the addresses of 21 hospitals in the county.

In the attribute table for the coverage `cuyatrt`, the item `age_under5` is the number of children under the age of 5, and the item `age_65_up` is the number of seniors above 65 years old. It is based on the 2000 Census. The hospital file `cuya_hosp.csv` is extracted from the Hospitals Names and Addresses at the Ohio Hospital Association website (http://www.ohanet.org/research/dataresources).

The planning problem is to locate five clinics among the 21 hospitals in order to serve the clients (children and seniors) in a most efficient way (i.e., minimizing total distance or time). The MINDISTANCE model in ArcInfo will be used to find the optimal solution. In practice, demand and candidate centers can be represented by points, polygons, or nodes on the network. When points or polygons are used, travel impedance is simply measured as Euclidean distance. When a network is used, travel impedance is measured as network distance or time. The method for point-based analysis is almost identical to that for polygon-based analysis. The following illustrates a polygon-based approach and a network-based approach.

10.4.1 PART 1: POLYGON-BASED ANALYSIS

The location-allocation commands in ArcInfo assume demand and supply (candidate or center) locations on the same coverage. In the polygon-based approach, demand and candidate locations are saved in the attribute table (PAT) for the polygon coverage. For this case study, hospitals are assumed to be located at their nearest tract centroids since the distances between them are minimal and negligible,[7] and both the demand and candidate information will be stored in the polygon coverage `cuyatrt`.

1. *Geocoding the hospitals*: In ArcCatalog, create a geocoding service by choosing Address Locators > Create New Address Locator > selecting U.S. Streets with Zone (File) > naming the new address locator `cuyahoga`;

under Primary table, Reference data, choose `tgr390351ka`; other
default values are okay.

Match addresses in ArcMap by choosing Tools > Geocoding > Geocode
Address. Select `cuyahoga` as the address locator, choose `cuya_hosp.csv`
as the address table, and save the result as a shapefile `hosp0`. Project the
shapefile to `hosp` using the projection file defined in `cuyatrt`.

2. *Defining demand*: Open the attribute table of `cuyatrt`, add a new item
   `patient`, and calculate it as `patient = age_under5 + age_65_up`,
   which defines the total demand in each tract.[8]

3. *Linking hospitals to nearest tracts*: This is done by four steps in ArcGIS:
   a. Use the "feature to point" tool to generate a point layer `cuyatrtpt`
      from `cuyatrt`.
   b. Use the "near" tool to find the nearest tract to each hospital by choosing
      `hosp` as the input feature and `cuyatrtpt` as the near feature.
   c. Join the attribute table of `cuyatrtpt` to `hosp` so that the attribute
      table of `hosp` contains additional identifications of the nearest tracts
      (e.g., `STFID`),[9] and export the combined table to a new table `hosptrt`.
   d. Join the table `hosptrt` to the tract coverage `cuyatrt` (using the
      common key `STFID`).

   Note that the result from the "near" tool shows that two hospitals
   (Rainbow Babies and Children's Hospital and University Hospitals of
   Cleveland) are close to each other and share the same nearest tract. This
   implies that the two hospitals will serve as only one candidate site for
   locating clinics. It reduces the candidate sites from 21 to 20.

4. *Defining candidates for site selection*: Add a new item `clinic` to the
   attribute table of `cuyatrt`, select the records with the criterion
   `hosp:id>0` (i.e., only those nearest tracts to hospitals), and calculate it
   as `clinic = 1`. The item `clinic` is the *candidate item* (= 0, not a
   candidate; = 1, a mobile candidate for site selection; = 2, a fixed candidate
   for site selection).

5. *Executing the location-allocation analysis*: In the ArcInfo Workstation
   environment, navigate to the project directory and activate ArcPlot by
   typing `ap` in the ArcInfo command line. Then type the following commands in ArcPlot prompt:

```
locatecandidates cuyatrt poly patient clinic

locatecriteria mindistance

locateallocate outalloc1 outcent1 outglob1 5
```

The command `locatecandidates` sets up the environment for analysis with (1) `cuyatrt` defining the coverage name, (2) `poly` identifying
the coverage type "polygon" instead of "point," (3) `patient` specifying
the demand item, and (4) `clinic` specifying the candidate item. The
command `locatecriteria` defines the model as MINDISTANCE. The
command `locateallocate` executes the model with (1) the INFO file
`outalloc1` containing a list of all demand locations, the amount of
demand, and the first and second closest centers and their distances; (2) the

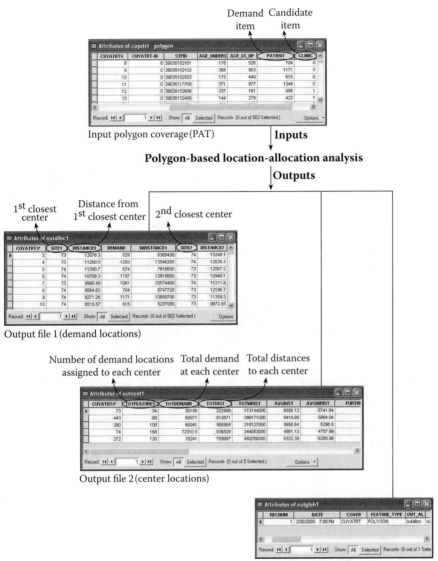

FIGURE 10.2 Input and output files in the polygon-based location-allocation analysis.

INFO file `outcent1` containing a list of chosen candidate centers and related information; (3) the INFO file `outglob1` containing a summary statement; and (4) **5** as the number of sites to locate.

Figure 10.2 illustrates the input and output files in the polygon-based location-allocation analysis.

6. *Displaying the analysis results*: Join both the tables `outalloc1` and `outcent1` to `cuyatrt` (based on the common key `cuyatrt#`). The item `site1` in `outalloc1` identifies the clinic a tract has been assigned

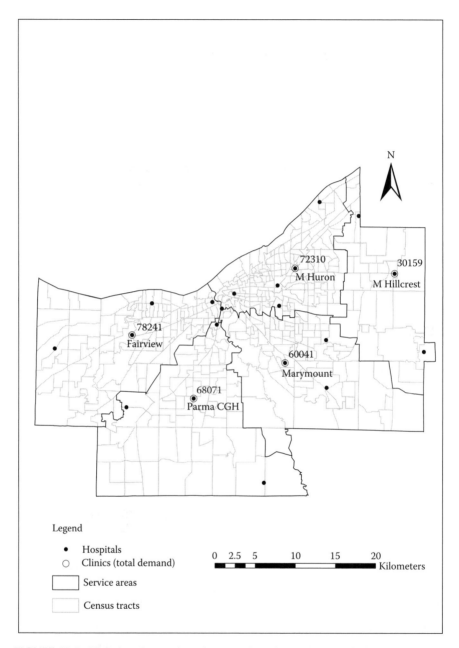

FIGURE 10.3 Clinic locations and service areas by polygon-based analysis.

to, based on which the coverage `cuyatrt` can be dissolved to five service areas for the five selected clinics. The item `totdemand` in `outcent1` is the aggregated demand (based on the item "patient") of tracts within each service area (see Figure 10.3). The results are also summarized in Table 10.2.

TABLE 10.2
Location-Allocation Analysis Results (Polygon Based vs. Network Based)

Clinic Locations	Service Areas by Polygon-Based Analysis		Service Areas by Network-Based Analysis	
	Tracts	Total Demand	Tracts	Total Demand
Meridia Hillcrest Hospital	34	30,159	32	28,428
Meridia Huron Hospital	168	72,311	168	79,005
Fairview Hospital	120	78,241	109	75,110
Marymount Hospital	100	60,041	96	55,064
Parma Community General Hospital	80	68,071	92	71,213

10.4.2 PART 2: NETWORK-BASED ANALYSIS

Similar to the polygon-based analysis, both the demand and candidate locations need to be saved in the network coverage for the network-based analysis. For this case study, demand information will be stored as an item on the node attribute table (NAT) of the network coverage, and candidate information will be stored as an item in the CENTERS file (using the same identification numbers as in the network NAT file). This case study uses network distance to measure travel impedance.

1. *Prepare the road coverage and others*: In ArcToolbox, (1) project the shapefile `tgr390351ka` to `cuyard0` using the projection file defined in `cuyatrt`, (2) use Conversion Tools > To Coverage > Feature Class To Coverage to convert the shapefile `cuyard0` to coverage `cuyard`, and (3) use Coverage Tools > Data Management > Topology > Build to build both line and node topologies on `cuyard`. One may view the line feature and the node feature of `cuyard` (as well as the corresponding attribute tables `cuyard.aat` and `cuyard.nat`) by adding each to the project in ArcMap.[10]

2. *Define the INFO file for candidate centers*: Similar to step 3 in Section 10.4.1, this step is to assign hospitals to the nearest nodes and then define the candidate locations. In ArcGIS, (1) use the "near" tool to find the nearest node to each hospital by choosing `hosp` as the input feature and `cuyard` (node) as the near feature, (2) join the *node* attribute table of `cuyard` to `hosp` so that the attribute table of `hosp` contains additional identifications of the nearest nodes, and export the combined table to a new table `hospnode.dbf`, and (3) add a field `clinic` in the table `hospnode.dbf` and calculate as `clinic = 1`. Unlike Part 1, the result from the "near" tool shows that each hospital corresponds to one unique nearest node, and thus 21 candidate locations.
It is necessary to access the ArcInfo Workstation for defining an INFO file. In the ArcInfo Workstation environment, (1) type the command `dbaseinfo hospnode.dbf centers` to convert the dBase file `hospnode.dbf` to INFO file `centers`, (2) rename the items

cuyard_ and cuyard_id in the INFO file centers to cuyard# and cuyard-id, respectively, and (3) drop all other items but three (cuyard#, cuyard-id, and clinic).

3. *Define the demands in the network NAT file*: Similar to step 2 above, this step is to link census tracts to the nearest nodes and then assign the demand information associated with the tracts to their nearest nodes. In ArcGIS, (1) use the "near" tool to find the nearest node to each tract by choosing cuyatrtpt as the input feature and cuyard (node) as the near feature,[11] and (2) join the *node* attribute table of cuyard to cuyatrtpt so that the attribute table of cuyatrtpt contains additional identifications of the nearest nodes, and export the combined table to a new table trtnode.dbf.

    Similarly, in the ArcInfo Workstation environment, (1) type the command dbaseinfo trtnode.dbf demand to convert the dBase file trtnode.dbf to INFO file demand, (2) rename the items cuyard_ and cuyard_id in the INFO file demand to cuyard# and cuyard-id, respectively, (3) drop all other items but four (cuyard#, cuyard-id, STFID, and patient), and (4) type the command joinitem cuyard.nat demand cuyard.nat cuyard# to join the INFO file demand to cuyard.nat.

4. *Executing the location-allocation analysis*: Activate ArcPlot by typing ap in the ArcInfo command line, and then type the following commands in ArcPlot prompt:

```
netcover cuyard cuyaroute

demand # patient

centers centers

locatecandidates centers clinic

locatecriteria mindistance

locateallocate outalloc2 outcent2 outglob2 5
```

    The command netcover uses cuyard as the network coverage to define the route system cuyaroute. The command demand skips the option "arc_demand_item" to "node_demand_item" defined by the item patient. The command centers uses the INFO file centers to define the candidate locations. The command locatecandidates centers sets up the analysis environment by using the candidate item clinic in the CENTERS file. Other commands (locatecriteria and locateallocate) are the same as in the polygon-based analysis. Figure 10.4 illustrates the input and output files in the network-based location-allocation analysis.

5. *Displaying the analysis results*: In ArcMap, (1) join both the tables outalloc2 and outcent2 to the INFO table demand (based on the common key cuyard#) and export the combined table to an external file

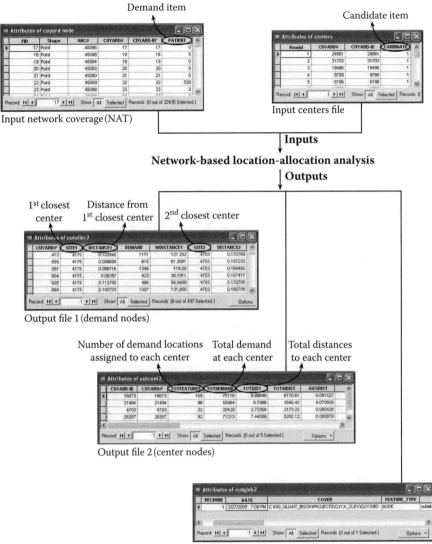

FIGURE 10.4 Input and output files in the network-based location-allocation analysis.

demand.dbf, (2) join the table demand.dbf to the coverage cuyatrt (based on the common key STFID), and (3) similarly, dissolve the coverage cuyatrt based on the item site1 to generate service areas for the five clinics. See Figure 10.5. The results are also summarized in Table 10.2.

Comparison between Figure 10.3 and Figure 10.5 reveals that both approaches site the five clinics at the same locations, but service areas by the network-based

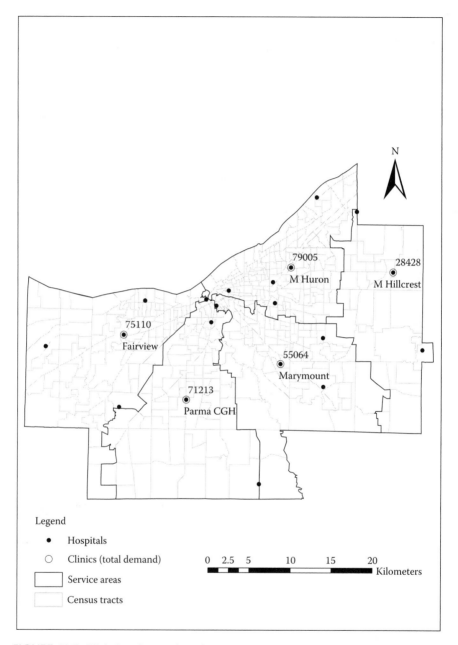

FIGURE 10.5 Clinic locations and service areas by network-based analysis.

analysis are slightly different from those by the polygon-based analysis because of the layout of the road network. For example, Interstate Highway I-90 (Figure 10.6) along the lakeshore makes the area at the county's northeast corner more accessible to Meridia Huron Hospital instead of Meridia Hillcrest Hospital. The State Route SR-91

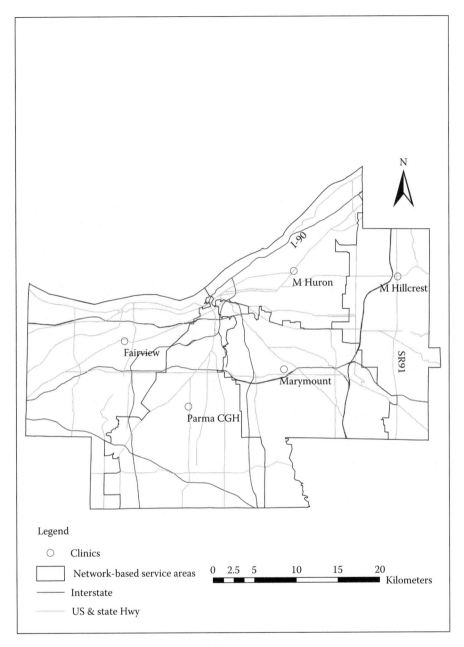

FIGURE 10.6 Highways in Cuyahoga, Ohio.

(Center Road) on the east side makes the area at the county's southeast corner more accessible to Meridia Hillcrest Hospital instead of Marymount Hospital. For the same reason, the service area of the Parma Community General Hospital expands toward the north but shrinks at both its southwest and southeast corners.

10.5 DISCUSSION AND SUMMARY

Applications of linear programming (LP) are widely seen in operational research, engineering, socioeconomic planning, location analysis, and others. This chapter first discusses the simplex algorithm and the implementation of solving LP problems in SAS. The LP procedure, available in the SAS/OR module, is a general tool for solving LP, integer LP, and mixed-integer LP problems. Any integer variables may be defined in the TYPE statement as INTEGER or BINARY in the LP procedure (though only non-integer variables are involved in the case study of measuring wasteful commuting). Specifically, the sparse format is recommended for coding the LP problems that involve large matrices. SAS/OR also has other procedures, such as ASSIGN (for assignment problems), CPM (for scheduling projects), GANTT (for plotting Gantt charts), NETDRAW (for drawing network diagrams), NETFLOW (for network flow programming), and TRANS (for transportation problems), to solve more specialized LP problems efficiently.

The wasteful commuting issue is used as a case study to illustrate the LP implementation in SAS. The solution to minimizing total commute in a city is the optimal (required or minimum) commute. In the optimal city, many journey-to-work trips are intrazonal (i.e., residence and workplace are in the same zone) and yield very little required commute, and thus a high percentage of wasteful commuting. Computation of intrazonal travel distance or time is particularly problematic, as the lengths of intrazonal trips vary. Unless the journey-to-work data are down to street addresses, it is difficult to measure commute distances or times accurately. The interest in examining wasteful commuting, to some degree, stimulates studies on commuting and issues related to it. One area with important implications in public policy is the issue of spatial mismatch and social justice (see Kain, 2004, for a recent review).

The location-allocation models are just a set of examples of LP applications in location analysis. The second case study in this chapter illustrates how one of the models (p-median problem) is applied for site selections of health care providers, and how the problem is solved by using some built-in models in ArcInfo. The polygon-based (or point-based) approach simply uses Euclidean distances between demand and supply locations and is fairly easy to implement in ArcInfo. The network-based approach uses travel distances or times through a road network and better captures travelers' behavior, and its implementation requires the preparation of a network coverage. In both approaches, demand and supply location and attribute information need to be saved on the same coverage. In the polygon-based (point-based) approach, this can be done either by combining the coverage of supply locations with that of demand locations to form one coverage or by finding the nearest demand locations from the supply locations and using those demand locations to represent the supply locations approximately (as adopted in case study 10B, Part 1). In the network-based approach, demand and supply information is assigned to their nearest nodes on the network. The convenience of using ArcInfo to solve location-allocation models is evident, as the results are ready to feed back to ArcGIS for mapping. However, its current capacity remains limited, as it has yet to incorporate complex constraints. Solving more advanced models requires programming in SAS or other specialized software packages.

APPENDIX 10A: HAMILTON'S MODEL ON WASTEFUL COMMUTING

Economists often make assumptions in order to build a model with reasonable complexity while capturing the most important essence of real-world issues. Like the monocentric urban economic model, Hamilton (1982) made some assumptions for a simplified urban structure. One also needs to note two possible limitations to Hamilton in the early 1980s: the lack of intraurban employment distribution data at a fine geographic resolution and the GIS technology at its developmental stage.

First, consider the commuting pattern in a monocentric city where all employment is concentrated at the CBD (or the city center). Assume that population is distributed according to a density function $P(x)$, where x is the distance from the city center. The concentric ring at distance x has an area size $2xdx$, and thus population $2xP(x)dx$, who travel a distance x to the CBD. Therefore, the total distance D traveled by a total population N in the city is the aggregation over the whole urban circle with a radius R:

$$D = \int_0^R x(2\pi xP(x))dx = 2\pi \int_0^R x^2 P(x)dx$$

Therefore, the average commute distance per person A is

$$A = \frac{D}{N} = \frac{2\pi}{N} \int_0^R x^2 P(x)dx \qquad (A10.1)$$

Now assume that the employment distribution is decentralized across the whole city according to a function $E(x)$. Hamilton believed that this decentralized employment pattern was more realistic than the monocentric one. Similar to Equation A10.1, the average distance of employment from the CBD is

$$B = \frac{2\pi}{J} \int_0^R x^2 E(x)dx \qquad (A10.2)$$

where J is the total number of employment in the city.

Assuming that residents can freely swap houses in order to minimize commute, the planning problem here is to minimize total commuting given the locations of houses and jobs. Note that employment is usually more centralized than population. The solution to the problem is that commuters always travel toward the CBD and stop at the nearest employer. Compared to a monocentric city, "displacement of a job from the CBD can save the worker a commute equal to the distance between the job and the CBD" (Hamilton, 1982, p. 1040). Therefore, optimal commute or *required commute or minimum commute* per person is the difference between the average distance of population from the CBD (A) and the average distance of employment from the CBD (B):

$$C = A - B = \frac{2\pi P_0}{N}\int_0^R x^2 e^{-tx}dx - \frac{2\pi E_0}{J}\int_0^R x^2 e^{-rx}dx \qquad (A10.3)$$

where both the population and employment density functions are assumed to be exponential, i.e., $P(x) = P_0 e^{-tx}$ (see Equation 6.1) and $E(x) = E_0 e^{-rx}$ respectively.

Solving Equation A10.3 yields

$$C = -\frac{2\pi P_0}{tN}R^{-2}e^{-tR} + \frac{2}{t} + \frac{2\pi E_0}{rJ}R^{-2}e^{-rR} - \frac{2}{r}$$

Hamilton studied 14 American cities of various sizes and found that the required commute only accounts for 13% of the actual commute, and the remaining 87% is wasteful. He further calibrated a model in which households choose their homes and job sites at random, and found that the random commute distances are only 25% over actual commuting distances, much closer than the optimal commute.

There are many reasons that people commute more than the required commute predicted by Hamilton's model. Some are recognized by Hamilton himself, such as bias in the model's estimations (residential and employment density functions, radial road network)[12] and the unrealistic assumptions. For example, residents do not necessarily move close to their workplaces when they change their jobs because of relocation costs and other concerns. Relocation costs are likely to be higher for homeowners than renters, and thus homeownership may affect commute. There are also families with more than one worker. Unless the jobs of all workers in the family are at the same location, it is impossible to optimize commute trips for each income earner. More importantly, residents choose their homes for factors not related to their job sites, such as accessibility to other activities (shopping, services, recreation and entertainment, etc.), quality of schools and public services, neighborhood safety, and others. Some of these factors are considered in research on explaining intraurban variation of actual commuting (e.g., Wang, 2001b; Shen, 2000).

APPENDIX 10B: SAS PROGRAM FOR THE LP PROBLEM OF MEASURING WASTEFUL COMMUTING

```
/*  LP.sas minimizes total commute time in Columbus, Ohio.
    By Fahui Wang on 4-14-2005                                    */

/*  Input the data of resident workers & jobs                     */
data study;
    infile 'c:\gis_quant_book\projects\columbus\urbtaz.txt';
    input taz $1-6 work emp; /*TAZ codes, # Workers, # jobs*/
```

```
    proc sort; by taz;
data work; set study (rename=(taz=tazr)); if work>0;
    oindex+1;        /*Create an index for origin TAZs      */
data emp; set study (rename=(taz=tazw)); if emp>0;
    dindex+1;        /*Create an index for destination TAZs      */
/* Input the data of O-D commute time                          */
data netdist0;
    infile 'c:\gis_quant_book\projects\columbus\odtime.txt';
    input tazr $1-6 tazw $9-14 @15 d; /*from_taz, to_taz, time*/
    proc sort; by tazw;
data netdist1;      /*attach index for destination taz        */
    merge emp(in=a) netdist0; by tazw; if a;
    proc sort; by tazr;
data netdist2;      /*attach index for origin taz            */
    merge work(in=a) netdist1; by tazr; if a;
    route=oindex*1000+dindex; /*Create unique code for a route*/
    proc sort; by route;

/* Build the LP model in sparse format                         */
data model;
    length _type_ $8 _row_ $8 _col_ $8;
    keep _type_ _row_ _col_ _coef_; /*four variables needed */
NI=812;   /*total number of origin TAZs */
NJ=931;   /*total number of destination TAZs */

/* Create the Constraints on Jobs */
Do j=1 to NJ; set emp;
/* 1st entry defines the upper bound (#jobs) for a TAZ         */
    _row_='EMP'||put(j,3.); /* Increase the space limit "3" if >999
        TAZs */
    _type_='LE';
    _col_=' RHS_';
    _coef_=emp;
    output;
/* the following defines variables & coefficients in the same row */
    _type_=' ';
    Do I=1 to NI;
        if emp~=. then do; /* for non-zero emp TAZs only       */
        _col_='X'||put(i,3.)||put(j,3.); /* Xij */
```

```
    _coef_=1.0; /* all coefficients are 1 */
    output;
    end;
  end;
end;

/* Create the Constraints on Resident Workers */
Do i=1 to Ni;
   set work;
   _row_='WRK'||put(i,3.);
   _type_='EQ';    /* All resident workers must be assigned */
                   /* Note total resident workers < total jobs */
   _col_='_RHS_';
   _coef_=work;
   output;
   _type_=' ';
   Do j=1 to Nj;
      if work~=. then do;
      _col_='X'||put(i,3.)||put(j,3.);
      _coef_=1.0;
      output;
      end;
   end;
end;

/* Create the objective function */
_row_='OBJ';
Do I=1 to NI;
   Do J= 1 to NJ;
       _type_=' ';
       set netdist2;
       if d~=. then do;
       _col_='X'||put(i,3.)||put(j,3.);
       _coef_=D;
       output;
       end;
   end;
end;
_type_='MIN';
```

```
_col_=' ';
_coef_=.;
output;

/*  Run the LP Problem */
proc lp sparsedata
    primalout=result noprint time=60000 maxit1=5000 maxit2=50000;
    reset time=60000 maxit1=5000 maxit2=50000;

data result; set result;
    if _value_>0 and _price_>0;
/* Save the result to an external file for  review                */
data _null_; set result;
    file 'c:\gis_quant_book\projects\columbus\junk.txt';
    put _var_ _value_;
/* convert the X variable to From_TAZ and To_TAZ index codes */
data junk;
    infile 'c:\gis_quant_book\projects\columbus\junk.txt';
    input oindex 2-4 dindex 5-7 Ncom ;
    route=oindex*1000+dindex;
    proc sort; by route;

/* attach the From_TAZ and To_TAZ codes, #commuters and time
   and save the result to an external file                       */
data f;
    merge netdist2 junk(in=a); by route; if a;
    file 'c:\gis_quant_book\projects\columbus\min_com.txt';
    put tazr $1-6 tazw $8-13 Ncom 15-21 @23 d 12.5;
run;
```

NOTES

1. The *dual problem* to the linear programming problem is $(-A^T, -c, -b)$. See Wu and Coppins (1981) for economic interpretations of a primary and a dual linear programming problem.
2. Use the link http://www-unix.mcs.anl.gov/otc/Guide/faq/linear-programming-faq.html to see a review of free and commercial programs for solving LP problems.
3. Field names such as `road#` and `road-id` are invalid in the ArcGIS desktop environment and have been automatically changed to `road_` and `road_id`, respectively in the dBase file `resid.dbf`. The item name `road_id` needs to be changed back to `road-id` for the NODEDISTANCE command to recognize it. This can be done in ArcCatalog (see step 4 in Section 2.3.2) or by using the commands in ArcInfo Workstation: (1) `TABLES`, (2) `SEL RESID`, (3) `ALTER ROAD_ID ROAD-ID 9 F` .

4. For simplicity, it is assumed that the travel speed on the air distance segments is 25 miles per hour.

5. _TYPE_ also takes the value INTEGER or BINARY in integer programming to identify variables being integer constrained or (0, 1) constrained, respectively.

6. Basically the intrazonal commute time simulated here is the sum of (1) time between a resident worker TAZ and its nearest node and (2) time between a job TAZ and its nearest node. Therefore, when the origin (resident worker) TAZ is the same as the destination (job) TAZ, the travel time is twice the time from a TAZ to its nearest node.

7. One may also create a coverage by combining the tract centroids and hospital locations and use it in a *point-based analysis*.

8. Any demand locations with zero demand (i.e., age_under5 + age_65_up=0 in this case) do not contribute to total travel distances (the objective function), and thus will not be assigned to a service provider in the following location-allocation analysis. For convenience of mapping results, one may code these tracts with patient = 0.01 (a nonzero small demand) prior to the analysis.

9. The only identification for tracts in the resulting table hosp from "near" is Near-FID, which is the FID in the point layer cuyatrtpt, not the polygon coverage cuyatrt.

10. In the Add Data dialog, double-click cuyard to display the various features (in this case, "arc," "node," and others) associated with the same coverage.

11. Again, the result here shows that each tract corresponds to one unique nearest node. If multiple tracts share one nearest node, demands need to be summed up and transferred to the nearest node.

12. Hamilton did not differentiate residents (population in general, including dependents) and resident workers (those actually in the labor force who commute). By doing so, it was assumed that the labor participation ratio was 100 percent across an urban area.

11 Solving a System of Linear Equations and Application in Simulating Urban Structure

This chapter introduces the method for solving a system of linear equations. The technique is used in many applications, including the popular input–output analysis (e.g., Hewings, 1985; see Appendix 11A for a brief introduction). Here, the method is illustrated in solving the Garin–Lowry model, a model widely used by urban planners and geographers for analyzing urban land use structure. A case study using a hypothetical city shows how the distributions of population and employment interact with each other and how the patterns can be affected by the transportation network. The GIS usage in the case study involves the computation of a travel time matrix and other data preparation tasks.

The method is fundamental in numerical analysis (NA) and is often used as a building block in many NA tasks, such as solving a system of nonlinear equations and the eigenvalue problem. Appendix 11B shows how the task of solving a system of linear equations is also imbedded in the method of solving a system of nonlinear equations.

11.1 SOLVING A SYSTEM OF LINEAR EQUATIONS

A system of n linear equations with n unknowns $x_1, x_2, \ldots, x_n$ is written as

$$\begin{cases} a_{11}x_1 + a_{12}x_2 + \ldots + a_{1n}x_n = b_1 \\ a_{21}x_1 + a_{22}x_2 + \ldots + a_{2n}x_n = b_2 \\ \ldots\ldots \\ a_{n1}x_1 + a_{n2}x_2 + \ldots + a_{nn}x_n = b_n \end{cases}$$

In the matrix form, it is

$$\begin{bmatrix} a_{11} & a_{12} & \cdots & a_{1n} \\ a_{21} & a_{22} & \cdots & a_{2n} \\ \vdots & \vdots & \ddots & \vdots \\ a_{n1} & a_{n2} & \cdots & a_{nn} \end{bmatrix} \begin{bmatrix} x_1 \\ x_2 \\ \vdots \\ x_n \end{bmatrix} = \begin{bmatrix} b_1 \\ b_2 \\ \vdots \\ b_n \end{bmatrix}$$

or simply

$$Ax = b \qquad\qquad (11.1)$$

If matrix A has a *diagonal* structure, Equation 11.1 becomes

$$
\begin{bmatrix}
a_{11} & 0 & \cdots & 0 \\
0 & a_{22} & \cdots & 0 \\
\vdots & \vdots & \ddots & \vdots \\
0 & 0 & \cdots & a_{nn}
\end{bmatrix}
\begin{bmatrix}
x_1 \\ x_2 \\ \vdots \\ x_n
\end{bmatrix}
=
\begin{bmatrix}
b_1 \\ b_2 \\ \vdots \\ b_n
\end{bmatrix}
$$

The solution is simple:

$$x_i = b_i \,/\, a_{ii}$$

If $a_{ii} = 0$ and $b_i = 0$, x_i can be any real number, and if $a_{ii} = 0$ and $b_i \neq 0$, there is no solution for the system.

There are two other simple systems with easy solutions. If matrix A has a *lower triangular* structure (i.e., all elements above the main diagonal are 0), Equation 11.1 becomes

$$
\begin{bmatrix}
a_{11} & 0 & \cdots & 0 \\
a_{21} & a_{22} & \cdots & 0 \\
\vdots & \vdots & \ddots & \vdots \\
a_{n1} & a_{n2} & \cdots & a_{nn}
\end{bmatrix}
\begin{bmatrix}
x_1 \\ x_2 \\ \vdots \\ x_n
\end{bmatrix}
=
\begin{bmatrix}
b_1 \\ b_2 \\ \vdots \\ b_n
\end{bmatrix}
$$

Assuming $a_{ii} \neq 0$ for all i, the *forward-substitution* algorithm is used to solve the system by obtaining x_1 from the first equation, substituting x_1 in the second equation to obtain x_2, and so on.

Similarly, if matrix A has an *upper triangular* structure (i.e., all elements below the main diagonal are 0), Equation 11.1 becomes

$$
\begin{bmatrix}
a_{11} & a_{12} & \cdots & a_{1n} \\
0 & a_{22} & \cdots & a_{2n} \\
\vdots & \vdots & \ddots & \vdots \\
0 & 0 & \cdots & a_{nn}
\end{bmatrix}
\begin{bmatrix}
x_1 \\ x_2 \\ \vdots \\ x_n
\end{bmatrix}
=
\begin{bmatrix}
b_1 \\ b_2 \\ \vdots \\ b_n
\end{bmatrix}
$$

The *back-substitution* algorithm is used to solve the system.

By converting Equation 11.1 to the simple systems as discussed above, one may obtain the solution for a general system of linear equations. Thus, if matrix A can be factored into the product of a lower triangular matrix L and an upper triangular matrix U, such as $A = LU$, Equation 11.1 can be solved in two stages:

1. $Lz = b$ solve for z
2. $Ux = z$ solve for x

The first one can be solved by the forward-substitution algorithm, and the second one by the back-substitution algorithm.

Among various algorithms for deriving the *LU factorization* (or *LU decomposition*) of A, one called *Gaussian elimination with scaled row pivoting* is used widely as an effective method. The algorithm consists of two steps: a factorization (or *forward-elimination*) phase and a solution (involving updating and back-substitution) phase (Kincaid and Cheney, 1991, p. 145). Computation routines for the algorithm of Gaussian elimination with scaled row pivoting can be found in various computer languages, such as FORTRAN (Press et al., 1992a), C (Press et al., 1992b), and C++ (Press et al., 2002). In the program SimuCity.for (see Appendix 11C), the FORTRAN subroutine LUDCOMP implements the first phase and the subroutine LUSOLVE implements the second phase. The two subroutines also call for two other simple routines, SCAL and AXPY. Free FORTRAN compilers can be downloaded from the website http://www.thefreecountry.com/compilers/fortran.shtml and others. The author used a free FORTRAN compiler g77 (free for downloading at http://www.gnu.org/software/fortran/fortran.html) for test running the programs. Section 11.3 discusses how the programs are utilized to solve the Garin–Lowry model. One may also use commercial software MATLAB (www.mathworks.com) or Mathematica (www.wolfram.com) for the task of solving a system of linear equations.

11.2 THE GARIN–LOWRY MODEL

11.2.1 BASIC VS. NONBASIC ECONOMIC ACTIVITIES

An interesting debate on the relation between population and employment distributions in a city is whether population follows employment (i.e., workers find residences near their workplaces to save commuting time) or vice versa (i.e., businesses locate near residents for recruiting workforce or providing services). The *Garin–Lowry model* (Lowry, 1964; Garin, 1966) argues that population and employment distributions interact with each other and are interdependent. However, different types of employment play different roles. The distribution of *basic employment* is independent of the population distribution pattern and may be considered exogenous. *Service (nonbasic) employment* follows population. On the other side, the population distribution is determined by the distribution patterns of both basic and service employment. See Figure 11.1 for illustration. The interactions between employment and population decline with distances, which are defined by a transportation network. Unlike the urban economic model built on the assumption of monocentric employment (see the Mills–Muth Economic Model in Appendix 6A), the Garin–Lowry model has the flexibility of simulating a population distribution pattern corresponding to any given basic employment pattern, and thus can be used to examine the

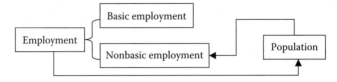

FIGURE 11.1 Interaction between population and employment distributions in a city.

impact of basic employment distribution on population as well as that of transportation network.

The binary division of employment into basic and service employment is based on the *concept of basic and nonbasic activities*. A local economy (a city or a region) can be divided into two sectors: basic sector and nonbasic sector. The *basic sector* refers to goods or services that are produced within the area but sold outside of the area. It is the export or surplus that is independent of the local economy. The *nonbasic sector* refers to goods or services that are produced within the area and also sold within the area. It is local or dependent and serves the local economy. By extension, basic employment refers to workers in the basic sector, and service employment refers to those in the nonbasic sector. The concept of basic and nonbasic activities is useful for several reasons (Wheeler et al., 1998, p. 140). It identifies the economic activities that are most important to a city's viability. Expansion or recession of the basic sector leads to economic repercussions throughout the city and affects the nonbasic sector. City and regional planners forecast the overall economic growth based on anticipated or predicted changes in the basic activities.

A common approach to determine employment in basic and nonbasic sectors is the *minimum requirements approach* by Ullman and Dacey (1962). The method examines many cities of approximately the same population size and computes the percentage of workers in a particular industry for each of the cities. If the lowest percentage represents the minimum requirements for that industry in a city of a given population-size range, that portion of the employment is engaged in the nonbasic or city-serving activities. Any portion beyond the minimum requirements is then classified as basic activity. Classifications of basic and nonbasic sectors can be also made by analyzing export data (Stabler and St. Louis, 1990).

11.2.2 THE MODEL'S FORMULATION

In the Garin–Lowry model, an urban area is composed of n tracts. The population in any tract j is affected by employment (including both the basic and service employment) in all n tracts, and the service employment in any tract i is determined by population in all n tracts. The degree of interaction declines with distance measured by a gravity kernel. Given a basic employment pattern and a distance matrix, the model computes the population and service employment at various locations.

First, the service employment in any tract i, S_i, is generated by the population in all tracts k ($k = 1, 2, ..., n$), P_k, through a gravity kernel t_{ik}, with

$$S_i = e \sum_{k=1}^{n} (P_k t_{ik}) = e \sum_{k=1}^{n} [P_k (d_{ik}^{-\alpha} / \sum_{j=1}^{n} d_{jk}^{-\alpha})] \tag{11.2}$$

where e is the service employment/population ratio (a simple scalar uniform across all tracts), d_{ik} the distance between tracts i and k, and α the distance friction coefficient characterizing shopping (resident-to-service) behavior. The gravity kernel t_{ik} represents the proportion of service employment in tract i owing to the influence of population in tract k, out of its impacts on all tracts. In other words, the service employment at i is a result of summed influences of population at all tracts k ($k = 1, 2, ..., n$), each of which is only a fraction of its influences on all tracts j ($j = 1, 2, ..., n$).

Second, the population in any tract j, P_j, is determined by the employment in all tracts i ($i = 1, 2, ..., n$), E_i, through a gravity kernel g_{ij}, with

$$P_j = h \sum_{i=1}^{n} (E_i g_{ij}) = h \sum_{i=1}^{n} [(B_i + S_i)(d_{ij}^{-\beta} / \sum_{k=1}^{n} d_{kj}^{-\beta})] \tag{11.3}$$

where h is the population/employment ratio (also a scalar uniform across all tracts) and β the distance friction coefficient characterizing commuting (resident-to-workplace) behavior. Note that employment E_i includes both service employment S_i and basic employment B_i, i.e., $E_i = S_i + B_i$. Similarly, the gravity kernel g_{ij} represents the proportion of population in tract j owing to the influence of employment in tract i, out of its impacts on all tracts k ($k = 1, 2, ..., n$).

Let P, S, and B be the vectors defined by the elements P_j, S_i, and B_i, respectively, and G and T the matrices defined by g_{ij} (with the constant h) and t_{ik} (with the constant e), respectively. Equations 11.2 and 11.3 become

$$S = TP \tag{11.4}$$

$$P = GS + GB \tag{11.5}$$

Combining Equations 11.4 and 11.5 and rearranging, we have

$$(I - GT)P = GB \tag{11.6}$$

where I is the $n \times n$ identity matrix. Equation 11.6 in the matrix form is a system of linear equations with the population vector P unknown. Four parameters (the distance friction coefficients α and β, the population/employment ratio h, and the service employment/population ratio e) are given; the distance matrix d is derived from a road network, and the basic employment B is predefined.

Plugging the solution P back to Equation 11.4 yields the service employment vector S. For more detailed discussion of the model, see Batty (1983).

The following subsection offers a simple example to illustrate the model.

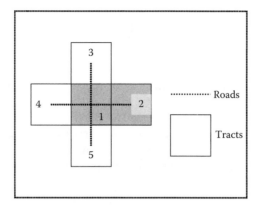

FIGURE 11.2 A simple city in the illustrative example.

11.2.3 AN ILLUSTRATIVE EXAMPLE

See Figure 11.2 for an urban area with five ($n = 5$) equal-area square tracts. The dashed lines are roads connecting them. Only two tracts (say, tracts 1 and 2, shaded in the figure) need to be differentiated, and carry different population and employment. Assume all basic employment is concentrated in tract 1 and *normalized* as 1, i.e., $B_1 = 1$, $B_2 = B_3 = B_4 = B_5 = 0$. This normalization implies that population and employment are relative, since we are only interested in their variation over space. The distance between tracts 1 and 2 is a unit 1, and the distance *within a tract* is defined as 0.25 (i.e., $d_{11} = d_{22} = d_{33} = \ldots = 0.25$). Note that the distance is the travel distance through the transportation network (e.g., $d_{23} = d_{21} + d_{13} = 1 + 1 = 2$). For illustration, define constants $e = 0.3$, $h = 2.0$, $\alpha = 1.0$, and $\beta = 1.0$.

From Equation 11.2, after taking advantage of the symmetric property (i.e., tracts 2, 3, 4, and 5 are equivalent in locations relative to tract 1), we have

$$S_1 = 0.3(\frac{d_{11}^{-1}}{d_{11}^{-1} + d_{21}^{-1} + d_{31}^{-1} + d_{41}^{-1} + d_{51}^{-1}}P_1 + \frac{d_{12}^{-1}}{d_{12}^{-1} + d_{22}^{-1} + d_{32}^{-1} + d_{42}^{-1} + d_{52}^{-1}}P_2 * 4)$$

That is,

$$S_1 = 0.1500P_1 + 0.1846P_2 \tag{11.7}$$

where the distances have been substituted by their values.

Similarly,

$$S_2 = 0.3(\frac{d_{21}^{-1}}{d_{11}^{-1} + d_{21}^{-1} + d_{31}^{-1} + d_{41}^{-1} + d_{51}^{-1}}P_1 + \frac{d_{22}^{-1}}{d_{12}^{-1} + d_{22}^{-1} + d_{32}^{-1} + d_{42}^{-1} + d_{52}^{-1}}P_2 +$$

$$\frac{d_{23}^{-1}}{d_{13}^{-1} + d_{23}^{-1} + d_{33}^{-1} + d_{43}^{-1} + d_{53}^{-1}} P_3 * 2 + \frac{d_{24}^{-1}}{d_{14}^{-1} + d_{24}^{-1} + d_{34}^{-1} + d_{44}^{-1} + d_{54}^{-1}} P_4)$$

where tracts 3 and 5 are equivalent in locations relative to tract 2. Noting $P_4 = P_3 = P_2$, the above equation is simplified as

$$S_2 = 0.0375P_1 + 0.2538P_2 \tag{11.8}$$

Similarly, from Equation 11.3 we have

$$P_1 = S_1 + 1.2308S_2 + 1 \tag{11.9}$$

$$P_2 = 0.2500S_1 + 1.6923S_2 + 0.25 \tag{11.10}$$

Solving the system of linear equations (Equations 11.7 to 11.10), we obtain $P_1 = 1.7472, P_2 = 0.8136; S_1 = 0.4123, S_2 = 0.2720$. Both the population and service employment are higher in the central tract than others.

11.3 CASE STUDY 11: SIMULATING POPULATION AND SERVICE EMPLOYMENT DISTRIBUTIONS IN A HYPOTHETICAL CITY

The hypothetical city is here assumed to be partitioned by a transportation network made of 10 contiguous circular rings and 15 radial roads. See Figure 11.3. Areas around the city center form a unique tract CBD, and thus the city has $1 + 9*15 = 136$ tracts. For convenience of network distance computation, we assume that each tract (except for the CBD, which is represented by the city center) enters or exits through the node intersected by the radial and the inner ring road. In other words, any non-CBD tracts are represented by these nodes on the road network. The hypothetical city does not have any geographic coordinate system or unit for distance measurement.

The following datasets are provided for the case study:

1. A polygon coverage `tract` contains 136 tracts of the city.
2. A road network coverage `road` is made of the same lines from the polygon coverage `tract`, but has the line and node topologies that have been built.
3. A point coverage `trtpt`, representing 135 non-CBD tracts, is extracted from the nodes contained in the road coverage `road`.
4. A point coverage `cbd` contains a single point for the location of CBD.

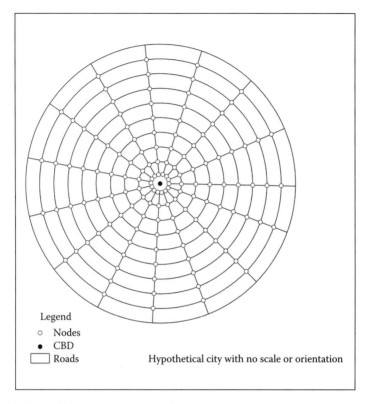

Legend
○ Nodes
● CBD
☐ Roads   Hypothetical city with no scale or orientation

FIGURE 11.3 Spatial structure of a hypothetical city.

The computation of road network distances (times) mainly utilizes the point coverage `trtpt` and the road network coverage `road`. The arc (line) attribute table for the road network coverage `road` (`road.aat`) contains a standard item `length`, which will be used to define the impedance values in the network travel distance computation. In addition, `road.aat` contains an additional item `length1`, which is defined as 1/2.5 of `length` for the seventh ring road, and the same as `length` for others. This item will be used to define the new impedance values when we examine the impact of a suburban beltway on the seventh ring road. For instance, when the travel speed on the beltway is assumed to be 2.5 times the speed on others, its travel time or impedance is 1/2.5 of others. The attribute table for the point coverage `trtpt` (or `cbd`) contains an item `trtid` identifying each tract and an item `trt_perim` as the perimeter of each tract. Tract perimeters will be used to calculate the average within-tract travel distances, approximated as 1/4 of the tract perimeters.

11.3.1 TASK 1: COMPUTING NETWORK DISTANCES (TIMES) IN ARCGIS

In the basic case (i.e., the reference case used to compare with others), travel speed is assumed to be uniform on all roads. Travel time is equivalent to travel distance, and the length on each road segment measures the travel impedance. Since the road network nodes are used to represent tract locations, the network travel distances are fairly easy to obtain. If necessary, refer to instructions in Section 2.3. The following provides a brief guideline:

1. Compute the network travel distances from the nodes defined in `trtpt` to the same nodes defined in `trtpt` through the road network `road`, and add the intratract travel distances at the origin and destination tracts to obtain the total travel distances between no-CBD tracts.
2. Compute the travel distances between the CBD tract and other non-CBD tracts (`trtpt`) as the Euclidean distances (travel distances through the radial roads are equivalent to the Euclidean distances), and add the intratract travel distances at the origin and destination tracts to obtain the total travel distances between them.
3. Compute the intratract travel distance within the CBD tract.

Output all distances to an external file `odtime.txt`, a space-separated text file containing $136 \times 136 = 18{,}496$ records with three variables: origin tract ID, destination tract ID, and distance between them. The file `odtime.txt` is sorted by the origin tract ID and the destination tract ID and saved as a space-delimited text file `odtime.prn`. For convenience, an AML program `rdtime.aml` for computing the travel distance (time) matrix and the data file `odtime.prn` are both enclosed in the CD for reference.

Repeat the task by using `length1` as the travel impedance values, and output the travel times to a similar text file `odtime1.txt`. Similarly, the file `odtime1.txt` is sorted by the origin tract ID and the destination tract ID, and saved as a space-delimited text file `odtime1.prn` (also enclosed in the CD). Note that in this case, travel impedance is defined as travel time instead of distance because the speed on the seventh ring road is faster than others.

11.3.2 TASK 2: SIMULATING DISTRIBUTIONS OF POPULATION AND SERVICE EMPLOYMENT IN THE BASIC CASE

The basic case, as in the monocentric model, assumes that all basic employment (say, 100) is concentrated at the CBD. In addition, the basic case assumes that $\alpha = 1.0$ and $\beta = 1.0$ for the two distance friction coefficients in the gravity kernels. The values of h and e in the model are set equal to 2.0 and 0.3, respectively, based on data from the *Statistical Abstract of the United States* (Bureau of Census, 1993). If P_T, B_T, and S_T are the total population and total basic and service employments, respectively, we have $S_T = eP_T$ and $P_T = hE_T = h(B_T + S_T)$, and thus $P_T = (h/(1 - he))B_T$. As B_T is normalized to 100, it follows that $P_T = 500$ and $S_T = 150$. Keeping h, e, and B_T constant throughout the analysis implies that the total population and employment (basic and service) remain constant. Our focus is on the effects of exogenous variations in the spatial distribution of basic employment and in the values of the travel friction parameters α and β, and on the impact of building a suburban beltway.

The FORTRAN program `simucity.for` in Appendix 11C (also enclosed in the CD) reads the travel distance (time) matrix `odtime.prn`, uses the

TABLE 11.1
Simulated Population and Service Employment Distributions in Various Scenarios

	Population				Service Employment			
Location	Basic Case[a]	Uniform Basic Employment	$\alpha,\beta = 2$	With a Suburban Beltway	Basic Case	Uniform Basic Employment	$\alpha,\beta = 2$	With a Suburban Beltway
1	8.3188	5.5322	16.8392	8.1525	1.8900	1.7046	3.4875	1.8197
2	7.1632	5.3157	12.3649	7.0032	1.8069	1.6356	3.1947	1.7391
3	5.2404	4.5684	6.2167	5.1022	1.5003	1.3923	1.8949	1.4407
4	4.2291	4.0251	3.9251	4.1066	1.2883	1.2171	1.2844	1.2346
5	3.5767	3.6616	2.7623	3.4981	1.1300	1.0985	0.9394	1.0948
6	3.1075	3.4179	2.0684	3.0942	1.0050	1.0177	0.7202	0.9979
7	2.7465	3.2629	1.6099	2.8178	0.9022	0.9649	0.5691	0.9323
8	2.4551	3.2398	1.2843	2.6641	0.8150	0.9522	0.4585	0.9069
9	2.2113	2.8620	1.0397	2.3538	0.7389	0.8407	0.3734	0.8012
10	2.0014	2.5659	0.8465	2.1094	0.6712	0.7533	0.3047	0.7181

[a]Basic case: All basic employment is concentrated at the CBD; $\alpha,\beta = 1$; travel speed is uniform on all roads.

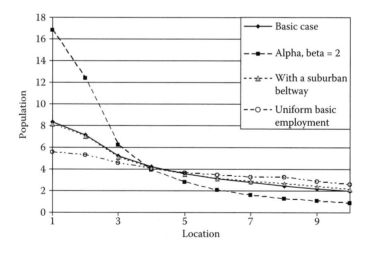

FIGURE 11.4 Population distributions in various scenarios.

LU decomposition method to solve the Garin–Lowry model, and outputs the results (numbers of population and service employment) to an external file `basic.txt`. Since the values are similar (or identical) for tracts on the same ring, 10 tracts from different rings along the same radial road (e.g., trtids = 11 to 19) are selected and shown in Table 11.1. Figure 11.4 and Figure 11.5 show the population and service employment patterns, respectively.

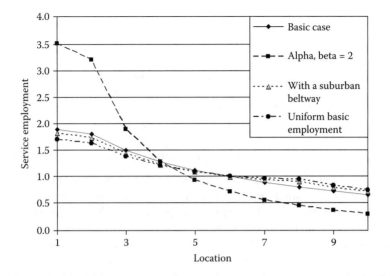

FIGURE 11.5 Service employment distributions in various scenarios.

11.3.3 TASK 3: EXAMINING THE IMPACT OF BASIC EMPLOYMENT PATTERN

To examine the impact of basic employment pattern, this project simulates the distributions of population and service employment given a uniform distribution of basic employment. In this case, all tracts have the same amount of basic employment, i.e., $100/136 = 0.8088$. In the program `simucity.for`, one only needs to modify the statements for coding the basic employment pattern and rerun it to obtain the result for a uniform basic employment pattern. The results are also shown in Table 11.1 and in Figure 11.4 and Figure 11.5. Note that both the population and service employment remain declining from the city center even when the basic employment is uniform across space. That is to say, the declining patterns with distances from the CBD are largely due to the location advantage (better accessibility) near the CBD shaped by the transportation network instead of job concentration in the CBD. The job concentration in CBD does enhance the effect; i.e., both the population and service employment exhibit flatter slopes in this case (uniform basic employment distribution) than in the basic case (a monocentric pattern). In general, service employment follows population, and their patterns are similar to each other.

One may design various basic employment patterns and examine how the population and service employment respond to the changes. See Guldmann and Wang (1998) for more scenarios of different basic employment patterns.

11.3.4 TASK 4: EXAMINING THE IMPACT OF TRAVEL FRICTION COEFFICIENT

Keep all parameters in the basic case unchanged except the two travel friction coefficients α and β. Compared to the basic case where $\alpha = 1$ and $\beta = 1$, this new case uses $\alpha = 2$ and $\beta = 2$. The travel friction parameters indicate how much people's travel behavior (including both commuting to workplace and shopping) is affected by travel distance (time). As transportation technologies as well as road networks improve over time, these two parameters have generally declined. In other words, the new case with $\alpha = 2$ and $\beta = 2$ may correspond to a city in earlier years.

Assigning new values to α and β in simucity.for yields new distribution patterns of population and service employment, also shown in Table 11.1 and in Figure 11.4 and Figure 11.5. Note the steeper slope in the case of larger α and β. That explains the flattening population density gradient over time, an important observation in the study of population density patterns (see Chapter 6).

11.3.5 Task 5: Examining the Impact of the Transportation Network

Finally, we examine the impact of the transportation network, in this particular case, the building of a suburban beltway. Assume that the seventh ring road is the faster beltway: if the average travel speed on other urban roads is 30 mph, the speed on the beltway is 75 mph. Task 1 has already generated a different travel time dataset, odtime1.prn, based on the new assumption. Changing the input file from odtime.prn to odtime1.prn in the program simucity.for yields the result under the new assumption, also shown in Table 11.1 and in Figure 11.4 and Figure 11.5. The distribution patterns of population and service employment are similar to those in the basic case, but both are slightly flatter than those in the basic case (note the lower values near the CBD and higher values near the 10th ring). In other words, building the suburban beltway narrows the gap in location advantage between the CBD and suburbia, and thus leads to flatter population and service employment patterns. See Wang (1998) for the impacts of different travel speeds and different locations of the suburban beltway.

11.4 DISCUSSION AND SUMMARY

The concept of basic and nonbasic activities emphasizes different roles in the economy played by basic and nonbasic sectors. The Garin–Lowry model uses the concept to characterize the interactions between employment and population distributions within a city. In the model, basic (export-oriented) employment serves as the exogenous factor, whereas service (locally oriented) employment depends on the population distribution pattern; on the other side, the distribution pattern of population is also determined by that of employment (including both basic and nonbasic employment). The interactions decline with travel distances or times as measured by gravity kernels. Based on this, the model is constructed as a system of linear equations. Given a basic employment pattern and a road network (the latter defines the matrix of travel distances or times), the model solves for the distributions of population and service employment.

Applying the Garin–Lowry model in analyzing real-world cities requires the division of employment into basic and nonbasic sectors. Various methods (e.g., the minimum requirements method) have been proposed to separate basic employment from the total employment. However, the division is unclear in many cases, as most economic activities in a city serve both the city itself (nonbasic sector) and beyond (basic sector). The case study uses a hypothetical city to illustrate the impacts of basic employment patterns, travel friction coefficients, and transportation networks. The case study helps us understand the change of urban structure under various scenarios and explain many empirical observations in urban density studies.

For example, suburbanization of basic employment in an urban area leads to dispersion of population as well as service employment, but the dispersion does not change the general pattern of higher concentrations of both population and employment toward the CBD. As the improvements in transportation technologies and road networks enable people to travel farther in less time, the traditional accessibility gap between the central city and suburbs is reduced and leads to a more gradual decline in population toward the edge of an urban area. This explains the flattening density gradient over time reported in many empirical studies on urban density functions. Suburban beltways "were originally intended primarily as a means of facilitating intercity travel by providing metropolitan bypass, it quickly became apparent that there were major unintended consequences for intracity traffic" (Taaffe et al., 1996, p. 178). The simulation in the case with a suburban beltway suggests a flatter population distribution pattern. In a study reported by Wang (1998, p. 274; with more ring roads and even faster speed on the beltway), the model even generates a suburban density peak near the beltway when all basic employment is assumed to be located in the CBD.

The model may be used to examine more issues in the study of urban structure. For example, comparing population patterns across cities with different numbers of ring roads sheds light on the issue of whether large cities exhibit flatter density gradients than smaller cities (McDonald, 1989, p. 380). Solving the model for a city with more radial or circular ring roads helps us to understand the impact of road density. Simulating cities with different road networks (e.g., a grid system, a semi-circular city) illustrates the impact of road network structure.

APPENDIX 11A: THE INPUT–OUTPUT MODEL

The *input–output model* is widely used in economic planning at various levels of governments. In the model, the output from any sector is also the input for all sectors (including the sector itself), and the inputs to one sector are provided by the outputs of all sectors (including itself). The key assumption in the model is that the *input–output coefficients* connecting all sectors characterize the technologies for a time period and remain unchanged over the period. The model is often used to examine how a change in production in one sector of the economy affects all other sectors, or how the productions of all sectors need to be adjusted in order to meet any changes in demands in the market.

We begin with a simple economy of two industrial sectors to illustrate the model. Consider an economy with two sectors and their production levels: X_1 for auto and X_2 for iron and steel. For each unit of output X_1, a_{11} is used as input (and thus a total amount of $a_{11}X_1$) in the auto industry itself; for each unit of output X_1, a_{12} is used as input (and thus a total amount of $a_{12}X_2$) in the iron and steel industry. In addition to inputs that are consumed within industries, d_1 serves the final demand to consumers. Similarly, X_2 has three components: $a_{21}X_1$ as the total input for the auto industry,

$a_{22}X_2$ as the total input for the iron and steel industry, and d_2 for the final demand. It is summarized as

$$\begin{cases} X_1 = a_{11}X_1 + a_{12}X_2 + d_1 \\ X_2 = a_{21}X_1 + a_{22}X_2 + d_2 \end{cases}$$

where the a_{ij} are the input–output coefficients.

In matrix,

$$IX = AX + D$$

where I is an identity matrix. Rearranging the equation yields

$$(I - A)X = D$$

which is a system of linear equations. Given any final demands in the future D and the input–output coefficients A, we can solve for the productions of all industrial sectors X.

APPENDIX 11B: SOLVING A SYSTEM OF NONLINEAR EQUATIONS

We begin with the solution of a single nonlinear equation by *Newton's method*. Say f is a nonlinear function whose zeros are to be determined numerically. Let r be a real solution and let x be an approximation to r. Keeping only the linear term in the *Taylor expansion*, we have

$$0 = f(r) = f(x + h) \approx f(x) + hf'(x) \tag{B11.1}$$

where h is a small increment, such as $h = r - x$. Therefore,

$$h \approx -f(x) / f'(x)$$

If x is an approximation to r, $x - f(x) / f'(x)$ should be a better approximation to r. Newton's method begins with an initial value x_0 and uses iterations to gradually improve the estimate until the function reaches an error criterion. The iteration is defined as

$$x_{n+1} = x_n - \frac{f(x_n)}{f'(x_n)}$$

The initial value assigned (x_0) is critical for the success of using Newton's method. It must be "sufficiently close" to the real solution (r) (Kincaid and Cheney,

1991, p. 65). Also, it only applies to a function whose first-order derivative has a definite form. The method can be extended to solve a system of nonlinear equations. Consider a system of two equations with two variables:

$$\begin{cases} f_1(x_1, x_2) = 0 \\ f_2(x_1, x_2) = 0 \end{cases} \tag{B11.2}$$

Similar to Equation B11.1, using the Taylor expansion, we have

$$\begin{cases} 0 = f_1(x_1 + h_1, x_2 + h_2) \approx f_1(x_1, x_2) + h_1 \frac{\partial f_1}{\partial x_1} + h_2 \frac{\partial f_1}{\partial x_2} \\ 0 = f_2(x_1 + h_1, x_2 + h_2) \approx f_2(x_1, x_2) + h_1 \frac{\partial f_2}{\partial x_1} + h_2 \frac{\partial f_2}{\partial x_2} \end{cases} \tag{B11.3}$$

This system of *linear* equations provides the basis for determining h_1 and h_2. The coefficient matrix is the *Jacobian matrix* of f_1 and f_2:

$$J = \begin{bmatrix} \frac{\partial f_1}{\partial x_1} & \frac{\partial f_1}{\partial x_2} \\ \frac{\partial f_2}{\partial x_1} & \frac{\partial f_2}{\partial x_2} \end{bmatrix}$$

Therefore, Newton's method for Equation B11.2 is

$$x_{1,n+1} = x_{1,n} + h_{1,n}$$

$$x_{2,n+1} = x_{2,n} + h_{2,n}$$

where the increments $h_{1,n}$ and $h_{2,n}$ are solutions to the rearranged system of linear equations (Equation B11.3):

$$\begin{cases} h_{1,n} \frac{\partial f_1}{\partial x_1} + h_{2,n} \frac{\partial f_1}{\partial x_2} = -f_1(x_{1,n}, x_{2,n}) \\ h_{1,n} \frac{\partial f_2}{\partial x_1} + h_{2,n} \frac{\partial f_2}{\partial x_2} = -f_2(x_{1,n}, x_{2,n}) \end{cases}$$

or

$$J \begin{bmatrix} h_{1,n} \\ h_{2,n} \end{bmatrix} = - \begin{bmatrix} f_1(x_{1,n}, x_{2,n}) \\ f_2(x_{1,n}, x_{2,n}) \end{bmatrix} \tag{B11.4}$$

Solving this system of linear equations uses the method discussed in Section 11.1. Solution of a larger system of nonlinear equations follows the same strategy, and only the Jacobian matrix is expanded. For instance, for a system of three equations, the Jacobian matrix is

$$J = \begin{bmatrix} \frac{\partial f_1}{\partial x_1} & \frac{\partial f_1}{\partial x_2} & \frac{\partial f_1}{\partial x_3} \\ \frac{\partial f_2}{\partial x_1} & \frac{\partial f_2}{\partial x_2} & \frac{\partial f_2}{\partial x_3} \\ \frac{\partial f_3}{\partial x_1} & \frac{\partial f_3}{\partial x_2} & \frac{\partial f_3}{\partial x_3} \end{bmatrix}$$

APPENDIX 11C: FORTRAN PROGRAM FOR SOLVING THE GARIN–LOWRY MODEL

```
*******************************************************************
*    SimuCity.FOR simulates the distributions of population      *
*    and service employment given a basic employment pattern     *
*    using the Garin-Lowry model: (I-A*B)*POPU=A*BEMP.           *
*    Version on 4/26/2005, by Fahui Wang.                         *
*******************************************************************

*    Variables defined:
*    ALPHA, BETA: distance friction coefficients
*    H: population / employment ratio
*    E: service employment / population ratio
*    N: total number of tracts the city is divided into
*    D(i,j): distance between tracts i and j
*    A(i,j), B(i,j): matrices G, T in the Garin-Lowry model
*    BEMP(i): basic employment vectors [known]
*    POP(i), SEMP(i): population & service employment vectors
*             [variables to be solved]
*    other variables: intermediates for computational purposes,
*             defined where they appear first

     PARAMETER (N=136)
     DOUBLE PRECISION D(N,N),A(N,N),B(N,N),DOMA(N),DOMB(N),IAB(N,N)
     DOUBLE PRECISION BEMP(N),POP(N),SEMP(N),AA(N,N),BB(N)
     REAL ALPHA,BETA,H,E
     INTEGER IPVT(N),INFO,I,J,OZONE(N),DZONE(N)
```

```
*      Step 1. Define parameters & Input data

*      Input the values of ALPHA, BETA, H, E
       DATA ALPHA /1.0/
       DATA BETA /1.0/
       DATA H /2.0/
       DATA E /0.3/

*      Input the distribution of Basic Employment
*      In the basic case, all employment (100) is assumed to be at CBD
       DO 1 I=1,N
1      BEMP(I)=0.0
       BEMP(1)=100.0

*      Input the distance matrix, pre-sorted by OZone & DZone
       OPEN (2, FILE='c:/g77/ODTIME.PRN', STATUS='OLD')
       DO 10 I=1,N
       DO 10 J=1,N
       READ (2,*) OZONE(I), DZONE(J), D(I,J)
10     CONTINUE
       CLOSE(2)

*      Step 2. Build matrices A, B, I-A*B, based on D(i,j)

*      Derive matrix Aij, Bij first

       DO 20 I=1,N
*           DOMA(i) & DOMB(i) are the dominators in matrix formula.
            DOMA(I)=0.0; DOMB(I)=0.0
       DO 20 J=1,N
            DOMA(I)=DOMA(I)+D(J,I)**(-BETA)
            DOMB(I)=DOMB(I)+D(J,I)**(-ALPHA)
20     CONTINUE
       DO 30 J=1,N
       DO 30 I=1,N
       A(J,I)=H*D(J,I)**(-BETA)/DOMA(I)
       B(J,I)=E*D(J,I)**(-ALPHA)/DOMB(I)
30     CONTINUE

*      Derive the matrix I-A*B, represented by IAB here
```

```
      DO 70 I=1,N
      DO 70 J=1,N
          IAB(I,J)=0.0
          DO 80 M=1,N
                IAB(I,J)=IAB(I,J)+A(I,M)*B(M,J)
80        CONTINUE
          IF (I.EQ.J) THEN
                IAB(I,J)=1.0-IAB(I,J)
          ELSE
                IAB(I,J)=-IAB(I,J)
          ENDIF
70    CONTINUE

*     Step 3. Prepare the data for solving the model.
*     Since it is a given-employment problem, in the
*     system of linear equations: AA*X=BB, AA is IAB,
*     BB is A*BEMP, and X is POP.

      DO 90 I=1,N
          BB(I)=0.0
          DO 90 J=1,N
                BB(I)=BB(I)+A(I,J)*BEMP(J)
                AA(I,J)=IAB(I,J)
90    CONTINUE

*     Step 4. Solve the system of linear equation, AA*X=BB

*     Factor AA matrix, print out INFO (indicating if the solution
*     exists).

      CALL LUDCOMP(AA,N,N,IPVT,INFO)
      PRINT *,INFO

*   Solve for x, which is vector POP

      CALL LUSOLVE(AA,BB,N,N,IPVT)

      DO 100 I=1,N
      POP(I)=BB(I)
      SEMP(I)=0.0
100   CONTINUE
```

```
*      Solve for the vetcor SEMP (=B*POP)
       DO 200 I=1,N
       DO 200 J=1,N
              SEMP(I)=SEMP(I)+B(I,J)*POP(J)
200    CONTINUE

*      Step 5. Output the results

       OPEN(12,FILE='Basic.TXT')
       DO 500 I=1,N
       WRITE(12,501) I,OZONE(I),BEMP(I),POP(I),SEMP(I)
501    FORMAT(1X,2(1x,i4),3(1X,f12.6))
500    CONTINUE
       CLOSE(12)

       STOP
       END

       subroutine scal(c,x,n)
c      ======================
c      Scales a vector by a constant
       integer i,n
       double precision c,x(*)
       do 10 i=1,n
              x(i) = c*x(i)
  10   continue
       return
       end

       subroutine axpy(c,x,y,n)
c      ========================
c      Constant times a vector plus a vector
       integer i,n
       double precision c,x(*),y(*)
       do 10 i=1,n
              y(i) = y(i) + c*x(i)
  10   continue
       return
       end
```

```
      subroutine ludcomp(a,n,m,ipvt,info)
c     ====================================
      integer m,n,ipvt(m),info
      double precision a(m,m)
c
c     ludcomp computes the L-U factors of a square
c     matrix by Gaussian elimination with pivoting
c
c     adapted from linpack
c
c     on entry
c
c     a       double precision(m,m)
c             matrix to be factored
c
c     n       integer
c             order of matrix a
c
c     m       integer
c     maximum order of matrix a
c
c     on return
c
c     a       upper triangular matrix U and, subdiagonally,
c             the multipliers in the L-U factorization of
c             the original matrix a
c
c     ipvt    integer(m)
c             vector of pivot indices
c
c     info    integer
c             = 0 normal value,
c             > 0 indicates singularity
c
c     calls: scal, axpy
c
      double precision t,dmax,dtmp
      integer ip,j,k
```

```
      info = 0
      do 60 k = 1, n-1
c             Find pivot index
              ip = k
              dmax = dabs(a(k,k))
              do 40 i = k+1,n
                    dtmp = dabs(a(i,k))
                    if(dtmp.le.dmax) go to 40
                    ip = i
                    dmax = dtmp
 40           continue
              ipvt(k) = ip
c             Zero pivot implies column already triangularized
              if (a(ip,k) .eq. 0.0d0) then
                    info = k
                    go to 60
              endif
c             Interchange if necessary
              if (ip .ne. k) then
                    t = a(ip,k)
                    a(ip,k) = a(k,k)
                    a(k,k) = t
              endif
c             Compute multipliers
              t = -1.0d0/a(k,k)
              call scal(t,a(k+1,k),n-k)
c             Row elimination with column indexing
              do 30 j = k+1, n
                    t = a(ip,j)
                    if (ip .ne. k) then
                          a(ip,j) = a(k,j)
                          a(k,j) = t
                    endif
                    call axpy(t,a(k+1,k),a(k+1,j),n-k)
 30           continue
 60   continue
      ipvt(n) = n
      if (a(n,n) .eq. 0.0d0) info = n
```

```
      return
      end

      subroutine lusolve(a,b,n,m,ipvt)
c     ================================
      integer n,ipvt(m)
      double precision a(m,m),b(m)
c
c     lusolve solves the double precision system
c     a * x = b using factors computed by ludcomp
c
c     adapted from linpack
c
c     on entry
c
c     a       double precision(m,m)
c             output from ludcomp
c
c     n       integer
c             order of matrix a
c
c     m       integer
c             maximum order of matrix a
c
c     ipvt    integer(m)
c             pivot vector from ludcomp
c
c     b       double precision(m)
c             right hand side vector
c
c     on return
c
c     b       solution vector x
c
c     error condition
c
c     division by zero will occur if the input factor contains a
c     zero on the diagonal; technically this indicates singularity
```

```
c
c     calls: axpy
c
      double precision t
      integer k,l

c     First solve l*y = b
      do 20 k = 1, n-1
            l = ipvt(k)
            t = b(l)
            if (l .eq. k) go to 10
                b(l) = b(k)
                b(k) = t
  10        continue
            call axpy(t,a(k+1,k),b(k+1),n-k)
  20  continue

c     Now solve u*x = y
      do 40 k = n, 1, -1
            b(k) = b(k)/a(k,k)
            t = -b(k)
            call axpy(t,a(1,k),b(1),k-1)
  40  continue

      return
      end
```

References

Abu-Lughod, J. 1969. Testing the theory of social area analysis: the ecology of Cairo, Egypt. *American Sociological Review* 34, 198–212.

Agnew, R. 1985. A revised strain theory of delinquency. *Social Forces* 64, 151–167.

Alonso, W. 1964. *Location and Land Use*. Cambridge, MA: Harvard University.

Alperovich, G. 1982. Density gradient and the identification of CBD. *Urban Studies* 19, 313–320.

Anderson, J.E. 1985. The changing structure of a city: temporal changes in cubic spline urban density patterns. *Journal of Regional Science* 25, 413–425.

Anselin, L. 1988. *Spatial Econometrics: Methods and Models*. Dordrecht, Netherlands: Kluwer.

Anselin, L. 1995. Local indicators of spatial association: LISA. *Geographical Analysis* 27, 93–115.

Anselin, L. and Bera, A. 1998. Spatial dependence in linear regression models with an introduction to spatial econometrics. In *Handbook of Applied Economic Statistics*, Ullah, A. and Giles, D.E., Eds. New York: Marcel Dekker, pp. 237–289.

Applebaum, W. 1966. Methods for determining store trade areas, market penetration and potential sales. *Journal of Marketing Research* 3, 127–141.

Applebaum, W. 1968. The analog method for estimating potential store sales. In *Guide to Store Location Research*, Kornblau, C., Ed. Reading, MA: Addison-Wesley.

Bailey, T.C. and Gatrell, A.C. 1995. *Interactive Spatial Data Analysis*. Harlow, England: Longman Scientific & Technical.

Baller, R., Anselin, L., Messner, S., Deane, G., and Hawkins, D. 2001. Structural covariates of U.S. county homicide rates: incorporating spatial effects. *Criminology* 39, 561–590.

Barkley, D.L., Henry, M.S., and Bao, S. 1996. Identifying "spread" versus "backwash" effects in regional economic areas: a density functions approach. *Land Economics* 72, 336–357.

Bartholdi, J.J., III and Platzman, L.K. 1988. Heuristics based on spacefilling curves for combinatorial problems in Euclidean space. *Management Science* 34, 291–305.

Batty, M. 1983. Linear urban models. *Papers of the Regional Science Association* 52, 141–158.

Batty, M. and Xie, Y. 1994a. Modeling inside GIS. Part I. Model structures, exploratory data analysis and aggregation. *International Journal of Geographical Information Systems* 8, 291–307.

Batty, M. and Xie, Y. 1994b. Modeling inside GIS. Part II. Selecting and calibrating urban models using arc-info. *International Journal of Geographical Information Systems* 8, 451–470.

Becker, G.S. 1968. Crime and punishment: an economic approach. *Journal of Political Economy* 76, 169–217.

Beckmann, M.J. 1971. On Thünen revisited: a neoclassical land use model. *Swedish Journal of Economics* 74, 1–7.

Bellair, P.E. and Roscigno, V.J. 2000. Local labor-market opportunity and adolescent delinquency. *Social Forces* 78, 1509–1538.

Berman, B. and Evans, J.R. 2001. *Retail Management: A Strategic Approach*, 8th ed. Upper Saddle River, NJ: Prentice Hall.

Berry, B.J.L. 1967. *The Geography of Market Centers and Retail Distribution*. Englewood Cliffs, NJ: Prentice Hall.

Berry, B.J.L. 1972. *City Classification Handbook, Methods, and Applications*. New York: Wiley-Interscience.

Berry, B.J.L. and Kim, H. 1993. Challenges to the monocentric model. *Geographical Analysis* 25, 1–4.

Berry, B.J.L. and Lamb, R. 1974. The delineation of urban spheres of influence: evaluation of an interaction model. *Regional Studies* 8, 185–190.

Berry, B.J.L. and Rees, P.H. 1969. The factorial ecology of Calcutta. *American Journal of Sociology* 74, 445–491.

Besag, J. and Newell, J. 1991. The detection of clusters in rare diseases. *Journal of the Royal Statistical Society Series A* 15, 4143–4155.

Black, R.J., Sharp, L., and Urquhart, J.D. 1996. Analysing the spatial distribution of disease using a method of constructing geographical areas of approximately equal population size. In *Methods for Investigating Localized Clustering of Disease*, Alexander, P.E. and Boyle, P., Eds. Lyon, France: International Agency for Research on Cancer, pp. 28–39.

Block, C.R., Block, R.L., and the Illinois Criminal Justice Information Authority (ICJIA). 1998. Homicides in Chicago, 1965–1995 [computer file], 4th ICPSR version. Chicago: ICJIA [producer]; Ann Arbor, MI: Inter-University Consortium for Political and Social Research [distributor].

Block, R. and Block, C.R. 1995. Space, place and crime: hot spot areas and hot places of liquor-related crime. In *Crime Places in Crime Theory*, Eck, J.E. and Weisburd, D., Eds. Newark, NJ: Criminal Justice Press.

Brabyn, L. and Gower, P. 2003. Mapping accessibility to general practitioners. In *Geographic Information Systems and Health Applications*, Khan, O. and Skinner, R., Eds. Hershey, PA: Idea Group Publishing, pp. 289–307.

Brown, B.B. 1980. Perspectives on social stress. In *Selye's Guide to Stress Research*, Vol. 1, Selye, H., Ed. Reinhold, NY: Van Nostrand, pp. 21–45.

Bureau of the Census. 1993. *Statistical Abstract of the United States*, 113th ed. Washington, DC: U.S. Department of Commerce.

Burgess, E. 1925. The growth of the city. In *The City*, Park, R., Burgess, E., and Mackenzie, R., Eds. Chicago: University of Chicago Press, pp. 47–62.

Cadwallader, M. 1975. A behavioral model of consumer spatial decision making. *Economic Geography* 51, 339–349.

Cadwallader, M. 1981. Towards a cognitive gravity model: the case of consumer spatial behavior. *Regional Studies* 15, 275–284.

Cadwallader, M. 1996. *Urban Geography: An Analytical Approach*. Upper Saddle River, NJ: Prentice Hall.

Casetti, E. 1993. Spatial analysis: perspectives and prospects. *Urban Geography* 14, 526–537.

Cervero, R. 1989. Jobs-housing balance and regional mobility. *Journal of the American Planning Association* 55, 136–150.

Chang, K.-T. 2004. *Introduction to Geographic Information Systems*, 2nd ed. New York: McGraw-Hill.

Chiricos, T.G. 1987. Rates of crime and unemployment: an analysis of aggregate research evidence. *Social Problems* 34, 187–211.

Christaller, W. 1966. *Central Places in Southern Germany*, Baskin, C.W., Trans. Englewood Cliffs, NJ: Prentice Hall.

Church, R.L. and ReVelle, C.S. 1974. The maximum covering location problem. *Papers of the Regional Science Association* 32, 101–118.

Clark, C. 1951. Urban population densities. *Journal of the Royal Statistical Society* 114, 490–494.

Clayton, D. and Kaldor, J. 1987. Empirical Bayes estimates of age-standardized relative risks for use in disease mapping. *Biometrics* 43, 671–681.

Cliff, A., Haggett, P., Ord, J., Bassett, K., and Davis, R. 1975. *Elements of Spatial Structure.* Cambridge, U.K.: Cambridge University.

Cliff, A.D. and Ord, J.K. 1973. *Spatial Autocorrelation.* London: Pion.

Colwell, P.F. 1982. Central place theory and the simple economic foundations of the gravity model. *Journal of Regional Science* 22, 541–546.

Cornish, D.B. and Clarke, R.V., Eds. 1986. *The Reasoning Criminal: Rational Choice Perspectives on Offending.* New York: Springer-Verlag.

Cressie, N. 1992. Smoothing regional maps using empirical Bayes predictors. *Geographical Analysis* 24, 75–95.

Cromley, E. and McLafferty, S. 2002. *GIS and Public Health.* New York: Guilford Press.

Curtin, K.M., Qiu, F., Hayslett-McCall, K., and Bray, T.M. 2005. Integrating Geographic Information Systems and maximal covering models to determine optimal police patrol force. In *GIS and Crime Analysis*, Wang, F., Ed. Hershey, PA: Idea Group Publishing, pp. 214–235.

Cuzick, J. and Edwards, R. 1990. Spatial clustering for inhomogeneous populations. *Journal of the Royal Statistical Society Series B* 52, 73–104.

Dantzig, G.B. 1948. *Programming in a Linear Structure.* Washington, DC: U.S. Air Force, Comptroller's Office.

Davies, R.L. 1973. Evaluation of retail store attributes and sales performance. *European Journal of Marketing* 7, 89–102.

Davies, W. and Herbert, D. 1993. *Communities within Cities: An Urban Geography.* London: Belhaven.

Diggle, P.J. and Chetwynd, A.D. 1991. Second-order analysis of spatial clustering for inhomogeneous populations. *Biometrics* 47, 1155–1163.

Dijkstra, E.W. 1959. A note on two problems in connection with graphs. *Numerische Mathematik* 1, 269–271.

Everitt, B.S., Landau, S., and Leese, M. 2001. *Cluster Analysis*, 4th ed. London: Arnold.

Fisch, O. 1991. A structural approach to the form of the population density function. *Geographical Analysis* 23, 261–275.

Flowerdew, R. and Green, M. 1992. Development in areal interpolation methods and GIS. *The Annals of Regional Science* 26, 67–78.

Forstall, R.L. and Greene, R.P. 1998. Defining job concentrations: the Los Angeles case. *Urban Geography* 18, 705–739.

Fotheringham, A.S., Brunsdon, C., and Charlton, M. 2000. *Quantitative Geography: Perspectives on Spatial Data Analysis.* London: Sage.

Fotheringham, A.S. and O'Kelly, M.E. 1989. *Spatial Interaction Models: Formulations and Applications.* London: Kluwer Academic.

Fotheringham, A.S. and Wong, D.W.S. 1991. The modifiable areal unit problem in multivariate statistical analysis. *Environment and Planning A* 23, 1025–1044.

Fotheringham, A.S. and Zhan, B. 1996. A comparison of three exploratory methods for cluster detection in spatial point patterns. *Geographical Analysis* 28, 200–218.

Franke, R. 1982. Smooth interpolation of scattered data by local thin plate splines. *Computers and Mathematics with Applications* 8, 273–281.

Frankena, M.W. 1978. A bias in estimating urban population density functions. *Journal of Urban Economics* 5, 35–45.

Gaile, G.L. 1980. The spread-backwash concept. *Regional Studies* 14, 15–25.

GAO. 1995. *Health Care Shortage Areas: Designation Not a Useful Tool for Directing Resources to the Underserved*, GAO/HEHS-95-2000. Washington, DC: General Accounting Office.

Garin, R.A. 1966. A matrix formulation of the Lowry model for intrametropolitan activity allocation. *Journal of the American Institute of Planners* 32, 361–364.

Geary, R. 1954. The contiguity ratio and statistical mapping. *The Incorporated Statistician* 5, 115–145.

Getis, A. 1995. Spatial filtering in a regression framework: experiments on regional inequality, government expenditure, and urban crime. In *New Directions in Spatial Econometrics*, Anselin, L. and Florax, R.J.G.M., Eds. Berlin: Springer, pp. 172–188.

Getis, A. and Griffith, D.A. 2002. Comparative spatial filtering in regression analysis. *Geographical Analysis* 34, 130–140.

Getis, A. and Ord, J.K. 1992. The analysis of spatial association by use of distance statistics. *Geographical Analysis* 24, 189–206.

Ghosh, A. and McLafferty, S. 1987. *Location Strategies for Retail and Service Firms*. Lexington, MA: D.C. Heath.

Giuliano, G. and Small, K.A. 1991. Subcenters in the Los Angeles region. *Regional Science and Urban Economics* 21, 163–182.

Giuliano, G. and Small, K.A. 1993. Is the journey to work explained by urban structure? *Urban Studies* 30, 1485–1500.

Goodchild, M.F., Anselin, L., and Deichmann, U. 1993. A framework for the interpolation of socioeconomic data. *Environment and Planning A* 25, 383–397.

Goodchild, M.F. and Lam, N.S.-N. 1980. Areal interpolation: a variant of the traditional spatial problem. *Geoprocessing* 1, 297–331.

Gordon, P., Richardson, H., and Wong, H. 1986. The distribution of population and employment in a polycentric city: the case of Los Angeles. *Environment and Planning A* 18, 161–173.

Greene, D.L. and Barnbrock, J. 1978. A note on problems in estimating exponential urban density models. *Journal of Urban Economics* 5, 285–290.

Griffith, D.A. 1981. Modelling urban population density in a multi-centered city. *Journal of Urban Economics* 9, 298–310.

Griffith, D.A. 2000. A linear regression solution to the spatial autocorrelation problem. *Journal of Geographical Systems* 2, 141–156.

Griffith, D.A. and Amrhein, C.G. 1997. *Multivariate Statistical Analysis for Geographers*. Upper Saddle River, NJ: Prentice Hall.

Grimson, R.C. and Rose, R.D. 1991. A versatile test for clustering and a proximity analysis of neurons. *Methods of Information in Medicine* 30, 299–303.

Gu, C., Wang, F., and Liu, G. 2005. The structure of social space in Beijing in 1998: a socialist city in transition. *Urban Geography* 26, 167–192.

Guldmann, J.M. and Wang, F. 1998. Population and employment density functions revisited: a spatial interaction approach. *Papers in Regional Science* 77, 189–211.

Haining, R., Wises, S., and Blake, M. 1994. Constructing regions for small area analysis: material deprivation and colorectal cancer. *Journal of Public Health Medicine* 16, 429–438.

Hamilton, B. 1982. Wasteful commuting. *Journal of Political Economy* 90, 1035–1053.

Hamilton, L.C. 1992. *Regression with Graphics*. Belmont, CA: Duxbury.

Hansen, W.G. 1959. How accessibility shapes land use. *Journal of the American Institute of Planners* 25, 73–76.

Harrell, A. and Gouvis, C. 1994. *Predicting Neighborhood Risk of Crime: Report to the National Institute of Justice*. Washington, DC: The Urban Institute.

Harris, C.D. and Ullman, E.L. 1945. The nature of cities. *The Annals of the American Academy of Political and Social Science* 242, 7–17.

Hartshorn, T.A. 1992. *Interpreting the City: An Urban Geography.* New York: John Wiley.

Heikkila, E.P., Gordon, P., Kim, J., Peiser, R., Richardson, H., and Dale-Johnson, D. 1989. What happened to the CBD-distance gradient? Land values in a polycentric city. *Environment and Planning A* 21, 221–232.

Hewings, G. 1985. *Regional Input-Output Analysis.* Beverly Hills, CA: Sage.

Hillsman, E. and Rushton, G. 1975. The *p*-median problem with maximum distance constraints. *Geographical Analysis* 7, 85–89.

Hirschi, T. 1969. *Causes of Delinquency.* Berkeley, CA: University of California Press.

Hoyt, H. 1939. *The Structure and Growth of Residential Neighborhoods in American Cities.* Washington, DC: USGPO.

Huff, D.L. 1963. A probabilistic analysis of shopping center trade areas. *Land Economics* 39, 81–90.

Huff, D.L. 2003. Parameter Estimation in the Huff Model. *ArcUser*, October/November, pp. 34–36.

Immergluck, D. 1998. Job proximity and the urban employment problem: do suitable nearby jobs improve neighborhood employment rates? *Urban Studies* 35, 7–23.

Jacquez, G.M. 1998. GIS as an enabling technology. In *GIS and Health*, Gatrell, A.C. and Loytonen, M., Eds. London: Taylor & Francis, pp. 17–28.

Jin, F., Wang, F., and Liu, Y. 2004. Geographic patterns of air passenger transport in China 1980–98: imprints of economic growth, regional inequality and network development. *Professional Geographer* 56, 471–487.

Joseph, A.E. and Bantock, P.R. 1982. Measuring potential physical accessibility to general practitioners in rural areas: a method and case study. *Social Science and Medicine* 16, 85–90.

Joseph, A.E. and Phillips, D.R. 1984. *Accessibility and Utilization: Geographical Perspectives on Health Care Delivery.* New York: Harper & Row.

Kain, J.F. 2004. A pioneer's perspective on the spatial mismatch literature. *Urban Studies* 41, 7–32.

Khan, A.A. 1992. An integrated approach to measuring potential spatial access to health care services. *Socio-economic Planning Science* 26, 275–287.

Khumawala, B.M. 1973. An efficient algorithm for the *p*-median problem with maximum distance constraints. *Geographical Analysis* 5, 309–321.

Kincaid, D. and Cheney, W. 1991. *Numerical Analysis: Mathematics of Scientific Computing.* Belmont, CA: Brooks/Cole Publishing Co.

Knox, P. 1987. *Urban Social Geography: An Introduction*, 2nd ed. New York: Longman.

Krige, D. 1966. Two-dimensional weighted moving average surfaces for ore evaluation. *Journal of South African Institute of Mining and Metallurgy* 66, 13–38.

Kulldorff, M. 1997. A spatial scan statistic. *Communications in Statistics: Theory and Methods* 26, 1481–1496.

Kulldorff, M. 1998. Statistical methods for spatial epidemiology: tests for randomness. In *GIS and Health*, Gatrell, A.C. and Loytonen, M., Eds. London: Taylor & Francis, pp. 49–62.

Ladd, H.F. and Wheaton, W. 1991. Causes and consequences of the changing urban form: introduction. *Regional Science and Urban Economics* 21, 157–162.

Lam, N.S.-N. and Liu, K. 1996. Use of space-filling curves in generating a national rural sampling frame for HIV-AIDS research. *Professional Geographer* 48, 321–332.

Land, K.C., McCall, P.L., and Cohen, L.E. 1990. Structural covariates of homicide rates: are there any in variances across time and social space? *American Journal of Sociology* 95, 922–963.

Land, K.C., McCall, P.L., and Nagin, D.S. 1996. A comparison of Poisson, negative binomial, and semiparametric mixed Poisson regression models: with empirical applications to criminal careers data. *Sociological Methods and Research* 24, 387–442.

Langford, I.H. 1994. Using empirical Bayes estimates in the geographical analysis of disease risk. *Area* 26,142–149.

Lee, R.C. 1991. Current approaches to shortage area designation. *Journal of Rural Health* 7, 437–450.

Levine, N. 2002. *CrimeStat: A Spatial Statistics Program for the Analysis of Crime Incident Locations, version 2.0.* Houston, TX: Ned Levine & Associates; Washington, DC: National Institute of Justice.

Levitt, S.D. 2001. Alternative strategies for identifying the link between unemployment and crime. *Journal of Quantitative Criminology* 17, 377–390.

Lindeberg, T. 1994. *Scale-Space Theory in Computer Vision.* Dordrecht, Netherlands: Kluwer Academic.

Lösch, A. 1954. *Economics of Location*, Woglom, W.H. and Stolper, W.F., Trans. New Haven, CT: Yale University.

Lowry, I.S. 1964. *A Model of Metropolis.* Santa Monica, CA: Rand Corporation.

Luo, J.-C., Zhou, C.-H., Leung, Y., Zhang, J.-S., and Huang, Y.-F. 2002. Scale-space theory based regionalization for spatial cells. *Acta Geographica Sinica* 57, 167–173 (in Chinese).

Luo, W. and Wang, F. 2003. Measures of spatial accessibility to healthcare in a GIS environment: synthesis and a case study in Chicago region. *Environment and Planning B: Planning and Design* 30, 865–884.

Marshall, R.J. 1991. Mapping disease and mortality rates using empirical Bayes estimators. *Applied Statistics* 40, 283–294.

McDonald, J.F. 1989. Econometric studies of urban population density: a survey. *Journal of Urban Economics* 26, 361–385.

McDonald, J.F. and Prather, P. 1994. Suburban employment centers: the case of Chicago. *Urban Studies* 31, 201–218.

Messner, S.F., Anselin, L., Baller, R.D., Hawkins, D.F., Deane, G., and Tolnay, S.E. 1999. The spatial patterning of county homicide rates: an application of exploratory spatial data analysis. *Journal of Quantitative Criminology* 15, 423–450.

Mills, E.S. 1972. *Studies in the Structure of the Urban Economy.* Baltimore: Johns Hopkins University.

Mills, E.S. and Tan, J.P. 1980. A comparison of urban population density functions in developed and developing countries. *Urban Studies* 17, 313–321.

Moran, P.A.P. 1950. Notes on continuous stochastic phenomena. *Biometrika* 37, 17–23.

Morenoff, J.D. and Sampson, R.J. 1997. Violent crime and the spatial dynamics of neighborhood transition: Chicago, 1970–1990. *Social Forces* 76, 31–64.

Muth, R. 1969. *Cities and Housing.* Chicago: University of Chicago.

Nakanishi, M. and Cooper, L.G. 1974. Parameter estimates for multiplicative competitive interaction models: least square approach. *Journal of Marketing Research* 11, 303–311.

Newling, B. 1969. The spatial variation of urban population densities. *Geographical Review* 59, 242–252.

Niedercorn, J.H. and Bechdolt, B.V., Jr. 1969. An economic derivation of the "gravity law" of spatial interaction. *Journal of Regional Science* 9, 273–282.

Oden, N., Jacquez, G., and Grimson, R. 1996. Realistic power simulations compare point- and area-based disease cluster tests. *Statistics in Medicine* 15, 783–806.

Olsen, L.M. and Lord, J.D. 1979. Market area characteristics and branch performance. *Journal of Bank Research* 10, 102–110.

Openshaw, S. 1984. *Concepts and Techniques in Modern Geography*, Number 38, *The Modifiable Areal Unit Problem*. Norwich: Geo Books.

Openshaw, S., Charlton, M., Mymer, C., and Craft, A.W. 1987. A Mark 1 geographical analysis machine for the automated analysis of point data sets. *International Journal of Geographical Information Systems* 1, 335–358.

Osgood, D.W. 2000. Poisson-based regression analysis of aggregate crime rates. *Journal of Quantitative Criminology* 16, 21–43.

Osgood, D.W. and Chambers, J.M. 2000. Social disorganization outside the metropolis: an analysis of rural youth violence. *Criminology* 38, 81–115.

Parr, J.B. 1985. A population density approach to regional spatial structure. *Urban Studies* 22, 289–303.

Parr, J.B. and O'Neill, G.J. 1989. Aspects of the lognormal function in the analysis of regional population distribution. *Environment and Planning A* 21, 961–973.

Parr, J.B., O'Neill, G.J., and Nairn, A.G.M. 1988. Metropolitan density functions: a further exploration. *Regional Science and Urban Economics* 18, 463–478.

Peng, Z. 1997. The jobs-housing balance and urban commuting. *Urban Studies* 34, 1215–1235.

Press, W.H. et al. 1992a. *Numerical Recipes in FORTRAN: The Art of Scientific Computing*, 2nd ed. Cambridge, U.K.: Cambridge University Press.

Press, W.H. et al. 1992b. *Numerical Recipes in C: The Art of Scientific Computing*, 2nd ed. Cambridge, U.K.: Cambridge University Press.

Press, W.H. et al. 2002. *Numerical Recipes in C++: The Art of Scientific Computing*, 2nd ed. Cambridge, U.K.: Cambridge University Press.

Price, M. 2004. *Mastering ArcGIS*. New York: McGraw-Hill.

Radke, J. and Mu, L. 2000. Spatial decomposition, modeling and mapping service regions to predict access to social programs. *Geographic Information Sciences* 6, 105–112.

Rees, P. 1970. Concepts of social space: toward an urban social geography. In *Geographic Perspectives on Urban System*, Berry, B. and Horton, F., Eds. Englewood Cliffs, NJ: Prentice Hall, pp. 306–394.

Reilly, W.J. 1931. *The Law of Retail Gravitation*. New York: Knickerbocker.

ReVelle, C.S. and Swain, R. 1970. Central facilities location. *Geographical Analysis* 2, 30–34.

Rogers, D.S. and Green, H. 1978. A new perspective in forecasting store sales: applying statistical models and techniques in the analog approach. *Geographical Review* 69, 449–458.

Rogerson, P.A. 1999. The detection of clusters using a spatial version of the chi-square goodness-of-fit statistic. *Geographical Analysis* 31, 130–147.

Rose, H.M. and McClain, P.D. 1990. *Race, Place, and Risk: Black Homicide in Urban America*. Albany, NY: SUNY Press.

Rushton, G. and Lolonis, P. 1996. Exploratory spatial analysis of birth defect rates in an urban population. *Statistics in Medicine* 7, 717–726.

Sampson, R.J., Raudenbush, S., and Earls, F. 1997. Neighborhoods and violent crime: a multilevel study of collective efficacy. *Science* 277, 918–924.

Shen, Q. 1994. An application of GIS to the measurement of spatial autocorrelation. *Computer, Environment and Urban Systems* 18, 167–191.

Shen, Q. 1998. Location characteristics of inner-city neighborhoods and employment accessibility of low-income workers. *Environment and Planning B: Planning and Design* 25, 345–365.

Shen, Q. 2000. Spatial and social dimensions of commuting. *Journal of the American Planning Association* 66, 68–82.

Sherratt, G. 1960. A model for general urban growth. In *Management Sciences: Models and Techniques*, Churchman, C.W. and Verhulst, M., Eds. Oxford: Pergamon Press.

Shevky, E. and Bell, W. 1955. *Social Area Analysis*. Stanford, CA: Stanford University.

Shevky, E. and Williams, M. 1949. *The Social Areas of Los Angeles*. Los Angeles: University of California.

Silverman, B.W. 1986. *Density Estimation for Statistics and Data Analysis*. London: Chapman & Hall.

Small, K.A. and Song, S. 1992. "Wasteful" commuting: a resolution. *Journal of Political Economy* 100, 888–898.

Small, K.A. and Song, S. 1994. Population and employment densities: structure and change. *Journal of Urban Economics* 36, 292–313.

Smirnov, O. and Anselin, L. 2001. Fast maximum likelihood estimation of very large spatial autoregressive models: a characteristic polynomial approach. *Computational Statistic and Data Analysis* 35, 301–319.

Stabler, J. and St. Louis, L. 1990. Embodies inputs and the classification of basic and nonbasic activity: implications for economic base and regional growth analysis. *Environment and Planning A* 22, 1667–1675.

Taaffe, E.J., Gauthier, H.L., and O'Kelly, M.E. 1996. *Geography of Transportation*, 2nd ed. Upper Saddle River, NJ: Prentice Hall.

Taneja, S. 1999. Technology Moves In. *Chain Store Age*, May, p. 136.

Tanner, J. 1961. Factors Affecting the Amount of Travel, Road Research Technical Paper 51. London: HMSO.

Tobler, W.R. 1970. A computer movie simulating urban growth in the Detroit region. *Economic Geography* 46, 234–240.

Toregas, C. and ReVelle, C.S. 1972. Optimal location under time or distance constraints. *Papers of the Regional Science Association* 28, 133–143.

Ullman, E.L. and Dacey, M. 1962. The minimum requirements approach to the urban economic base. *Proceedings of the IGU Symposium in Urban Geography*. Lund: Lund Studies in Geography, pp. 121–143.

Von Thünen, J.H. 1966. *Von Thünen's Isolated State*, Wartenberg, C.M., Trans.; Hall, P., Ed. Oxford: Pergamon.

Wang, F. 1998. Urban population distribution with various road networks: a simulation approach. *Environment and Planning B: Planning and Design* 25, 265–278.

Wang, F. 2000. Modeling commuting patterns in Chicago in a GIS environment: a job accessibility perspective. *Professional Geographer* 52, 120–133.

Wang, F. 2001a. Regional density functions and growth patterns in major plains of China 1982–90. *Papers in Regional Science* 80, 231–240.

Wang, F. 2001b. Explaining intraurban variations of commuting by job accessibility and workers' characteristics. *Environment and Planning B: Planning and Design* 28, 169–182.

Wang, F. 2003. Job proximity and accessibility for workers of various wage groups. *Urban Geography* 24, 253–271.

Wang, F. 2004. Spatial clusters of cancers in Illinois 1986–2000. *Journal of Medical Systems* 28, 237–256.

Wang, F. 2005. Job access and homicide patterns in Chicago: an analysis at multiple geographic levels based on scale-space theory. *Journal of Quantitative Criminology* 21, 195–217.

Wang, F. and Guldmann, J.M. 1996. Simulating urban population density with a gravity-based model. *Socio-economic Planning Sciences* 30, 245–256.

Wang, F. and Guldmann, J.M. 1997. A spatial equilibrium model for region size, urbanization ratio, and rural structure. *Environment and Planning A* 29, 929–941.

Wang, F. and Luo, W. 2005. Assessing spatial and nonspatial factors in healthcare access in Illinois: towards an integrated approach to defining health professional shortage areas. *Health and Place* 11, 131–146.

Wang, F. and Minor, W.W. 2002. Where the jobs are: employment access and crime patterns in Cleveland. *Annals of the Association of American Geographers* 92, 435–450.

Wang, F. and O'Brien, V. 2005. Constructing geographic areas for analysis of homicide in small populations: testing the herding-culture-of-honor proposition. In *GIS and Crime Analysis*, Wang, F., Ed. Hershey, PA: Idea Group Publishing, pp. 83–100.

Wang, F. and Zhou, Y. 1999. Modeling urban population densities in Beijing 1982–90: suburbanisation and its causes. *Urban Studies* 36, 271–287.

Webber, M.J. 1973. Equilibrium of location in an isolated state. *Environment and Planning A* 5, 751–759.

Weibull, J.W. 1976. An axiomatic approach to the measurement of accessibility. *Regional Science and Urban Economics* 6, 357–379.

Weisbrod, G.E., Parcells, R.J., and Kern, C. 1984. A disaggregate model for predicting shopping area market attraction. *Journal of Marketing* 60, 65–83.

Weisburd, D. and Green, L. 1995. Policing drug hot spots: the Jersey City drug market analysis experiment. *Justice Quarterly* 12, 711–735.

Wheeler, J.O. et al. 1998. *Economic Geography*, 3rd ed. New York: John Wiley & Sons.

White, M.J. 1988. Urban commuting journeys are not "wasteful." *Journal of Political Economy* 96, 1097–1110.

Whittemore, A.S., Friend, N., Brown, B.W., and Holly, E.A. 1987. A test to detect clusters of disease. *Biometrika* 74, 631–635.

Wilson, A.G. 1967. Statistical theory of spatial trip distribution models. *Transportation Research* 1, 253–269.

Wilson, A.G. 1974. *Urban and Regional Models in Geography and Planning*. London: Wiley.

Wilson, A.G. 1975. Some new forms of spatial interaction models: a review. *Transportation Research* 9, 167–179.

Wong, Y.-F. 1993. Clustering data by melting. *Neural Computation* 5, 89–104.

Wong, Y.-F. and Posner, E.C. 1993. A new clustering algorithm applicable to multispectral and polarimetric SAR images. *IEEE Transactions on Geoscience and Remote Sensing* 31, 634–644.

Wu, N. and Coppins, R. 1981. *Linear Programming and Extensions*. New York: McGraw-Hill.

Xie, Y. 1995. The overlaid network algorithms for areal interpolation problem. *Computer, Environment and Urban Systems* 19, 287–306.

Yang, Q. 1990. A model for interregional trip distribution in China. *Acta Geographica Sinica* (*Di-li Xue-bao*, in Chinese) 45, 264–274.

Zheng, X.-P. 1991. Metropolitan spatial structure and its determinants: a case study of Tokyo. *Urban Studies* 28, 87–104.

Zipf, G.K. 1949. *Human Behavior and the Principle of Least Effort*. Cambridge, MA: Addison-Welsey.

Index

A

Absolute errors, 105, 106
Accessibility index, 95
Accessibility, issues, 77–79
Accuracy, distance measurement, 19
Address matching, 65
Agglomerative hierarchical methods (AHMs), 132
Aggregate score, 153
Aggregated data, 161–162
AHMs, *see* Agglomerative hierarchical methods
Air distances, 27, 29, 30
Alonso's urban land use model, 99
AML, *see* Arc Micro Language
Analog method, 55, 56
Arc Micro Language (AML) program, 15, 28, 197, 227
ArcCatalog, 2, 3, *see also* ArcGIS
ArcGIS
    analyzing rate events in small populations, 151
    China
        point-based spatial cluster analysis Tai place-names, 171
        social area analysis, 140
    function fittings of polycentric models, 110
    gravity models for delineating trade areas, 57, 60
    importing/exporting ASCII files, 17–18
    measuring distances and time, 20
        network distance or time, 23–24
    Ohio
        extracting census tracts and analyzing polygon adjacency, 12–15
        mapping population density patterns, 4–8
        wasteful commuting, 194–197
    simulations of population/employment distributions in hypothetical city, 226–227
    spatial analysis tools, 8–12
    spatial and attribute data management, 1–4
    spatial smoothing and spatial interpolation, 36–37, 38, 43
    urban density patterns, 115, 116–117
ArcInfo system, 2, *see also* ArcGIS
ArcINFO Workstation
    allocating health care providers, 203, 204, 208
    measuring network distances or time in ArcGIS, 23

measuring travel distance between counties and major cities in China, 28
solving location-allocation problems, 201–202
ArcMap, 2, 3, 41, 65, 162, *see also* ArcGIS
ArcPlot, 29, 152, *see also* ArcGIS
Arcs, 22
ArcToolbox, 2, 65, 85, *see also* ArcGIS
Area-based methods, 167
Area-based spatial cluster analysis, 172–175, *see also* Cluster analysis
Areal weighting interpolator, 47, 116
ASCII file, 17–18
ASCII format, 3
Aspatial access, 78
Attraction-constrained model, 60
Attribute data, 7
    management, 1–4
Attribute file, 137
Attribute homogeneity, 150
Attribute join, 9, *see also* Join
Attribute queries, 8
Attribute table
    China
        analyzing Tai place-names by spatial smoothing, 39
        measuring distance between counties and major cities, 26, 29
        point-based spatial cluster analysis Tai place-names, 170
    data management in ArcGIS, 4
    defining fan bases, 65, 66
    Illinois
        analysis of urban density patterns, 115, 117
        measuring accessibility to primary care physicians, 90
        job access versus homicide patterns, 156, 161, 162
        spatial regression analysis of homicide patterns, 182
    Ohio
        allocating health care providers, 203, 204, 207
        measuring wasteful commuting, 194
    simulations of population and employment distributions in hypothetical city, 226
Average commute, 213
Average linkage method, 133

B

Back-substitution algorithm, 220
Backwash, 100, *see also* Centralization
Bandwidth, 37, 172
Basic case, 227–229, *see also* Hypothetical city
Basic economic activity, 221–222, *see also*
    GARIN–LOWRY model
Basic sector, 222, *see also* GARIN–LOWRY model
Basic variables, 192, 193
Bayesian inference, 52
Beijing, 135–143, *also* China
Bernoulli model, 169
β-values, 62–63, 84, 98, 99
Bias, 214
Bivariate functions, 101, 103, 113
Bivariate regression model, 62
BP, *see* Breaking point
Branch-and-bound method, 200
Breaking point (BP), 58, 59, 60–61
Breast cancer, 175, *see also* Cancer
Buffer, 10, *see also* Map overlays
Buffering tract, 14

C

CA, *see* Cluster analysis
Cancer, 175–180, *see also* Illinois
Cartesian coordinates, 19
Case files, 168, 170
Catchment area, 79
CBD, *see* Central business district
Census data, 5
Census Feature Class Code (CFCC), 23, 89
Census tract centroids, 112, 115
Census tracts, *see also* Centroids
    Illinois
        defining fan bases, 63
        job access versus homicide patterns, 158, 159
        measuring spatial accessibility to primary
          care physicians, 85
        spatial regression analysis of homicide
          patterns, 183–185
    Ohio
        analyzing polygon adjacency, 12–15
        area-based data aggregation of
          neighborhoods and school districts,
          48–51
Census Transportation Planning Package (CTPP),
    36, 48, 194, 199
Central business district (CBD)
    analysis of urban density patterns, 112, 117
    function fittings by regressions, 97, 119

simulating distributions of populations/service
    employment in hypothetical city, 225,
    229, 230
wasteful commuting, 193, 194, 213, 214
Central place theory, 57, 108
Centralization, 99, 100
Centroid method, 133
Centroids, 13–14, 24, 26, 85
CFCCs, *see* Census feature class codes
Changchun, 25, 68, 71, 72, *see also* China
Chicago
    job access versus homicide patterns, 155–163
    measuring spatial accessibility to primary care
      physicians, 84–91
    spatial cluster analysis of cancer patterns,
      175–180
    spatial regression analysis of homicide
      patterns, 182–185
    urban density pattern analysis, 110–117
Chicago consolidated metropolitan statistical area
    (CMSA), 63, 84, 111, *see also*
    Chicago; Illinois
Chicago Cubs, 63–68
Chicago school, 110–111
Chicago White Sox, 63–68
China
    analyzing Tai place-names by spatial smoothing,
      38–42
    defining hinterlands of major cities in northeast,
      68–71
    measuring distance between counties and
      major cities, 24–31
    spatial cluster analysis of Tai place-names in
      southern, 170–171
Clark's model, 97–98
Cleveland, *see also* Ohio
    analyzing polygon adjacency, 12–15
    area-based data aggregation, 48–51
Clip, 10, *see also* Map overlays
Cluster analysis (CA)
    multivariate statistical analysis method, 127,
      131–134
    social area analysis, 140, 142
Clustering tool, 158, 160
CMSA, *see* Chicago consolidated metropolitan
    statistical area
Cobb–Douglas form, 73
Colorectal cancer, 175, *see also* Cancer
Columbus MSA, 195, *see also* Ohio
Columbus, Ohio, 193–199
Comma-separated value file, 203
Common factors, 128
Community area, 185
Complete linkage method, 133

Computerized mapping, 1
Concentric zone model, 135
Connections, two-step, 32
Constraints, 198, 199, 202
Control file, 170
Control theory, 155
Controls, 168
Conversion tool, 5
Coordinates file, 170
Corner store sites, 56
Coverage, 194
Coverage data model, 3
Coverage interchange format, 5
Coverage road, 194, 196
Crime, 155–156, 167
CrimeStat, 168
Cross-area aggregation, *see* Spatial interpolation
CSV file, 17
CTPP, *see* Census Transportation Planning Package
Cubic spline function, 102, 105
Cubic trend surface model, 42
Cubs, *see* Chicago Cubs
Cutting planes method, 200
Cuyahoga County, Ohio, *see also* Ohio
    allocating health care providers, 203–211
    area-based data aggregation of school districts,
        49–51
    mapping population density pattern, 4–8, 9

D

Dalian, 25, 69, 71, *see also* China
Data management, 1–4
dBase file, 5, 31
dBase format, 3
dBase table, 27–28
Decentralization, 99, 100, 101
Dendrogram, 132
Dense format, 197, 198
Density crater, 102
Density functions, 97–101
Density patterns, 97
Department of Health and Human Services (DHHS),
        36, 79, 84
Destination layer, 10
Destination nodes, 27, *see also* Nodes
DFA, *see* Discriminant function analysis
DHHS, *see* Department of Health and Human
        Services
Diagonal (matrix) structure, 220
Dialog windows, 6
Directed network, 21, *see also* Network
Discriminant function analysis (DFA), 131, 145–146

Disease rates, 167
Distance, cluster analysis, 131
Distance decay effects, 119
Distance fraction coefficient, 60
Distance friction coefficient , *see* β-Values
Distance join, 10, 11, 20, *see also* Join
Distance matrix, 38–41, 66
Distance table, 26, 69, 86
Distance/time, measuring
    between counties and major cities in China,
        24–31
    computing network, 21–24
    measures, 19–21
    valued-graph approach to shortest-route
        problem, 31–32
Doubly constrained model, 60
Dummy variables, 135, 140–141, 142, 143

E

EB, *see* Empirical Bayes estimation
Economic model, 98
Economic theories, 155
Edge effects, 36, 93
Eigenfunction decomposition method, 188
Eigenvalue-1 rule, 129, 138
Eigenvalues, 129, 130, 138, 139
Empirical Bayes (EB) estimation, 52–53
Employment, 193, 194, 221–223, 229, *see also* Jobs;
        Wasteful commuting
Enumeration districts (EDs), 150
Ethnic status factor, 135, 136
Ethnicity, 140, 142, 144, 145
Euclidean distance
    allocating health care providers, 203
    area-based spatial cluster analysis, 172
    cluster analysis, 131
    defining fan bases, 66
    floating catchment area method in spatial
        smoothing, 36
    measuring distance between counties and
        major cities in China, 24–26, 28
    measuring distances and time, 19
    regression implementations and GIS, 102–103
Exponential functions, 113, 114, 117

F

FA, *see* Factor analysis
Factor analysis (FA)
    multivariate statistical analysis method, 127
    social area analysis, 127–131, 138

job access versus homicide patterns, 156, 162–163
spatial regression analysis of homicide patterns, 184, 185
Factor loadings, 129, 139
Factor score coefficients, 129
Factor scores, 129, 140, 142, 143
Factorial ecological approach, 135
False positive circles, 168
Family status factor, 134–135, 136
Family structure, 144–145
Fan bases, 63–68
Farthest-neighbor method, *see* Complete linkage method
FCA, *see* Floating catchment area
Field name, 3, 4
First-order conditions, 74, 120
Floating catchment area (FCA) method, 36–41
FORTRAN language
    simulating distributions of populations/service employment in hypothetical city, 227–228
    solving a system of linear equations, 221
    solving GARIN–LOWRY model, 234–241
Forward-substitution algorithm, 221
Functional fittings, regressions
    analyzing urban density patterns, 110–117
    density function approach to urban/regional structures, 97–101
    deriving urban density functions, 120–121
    monocentric models, 101–105
    nonlinear and weighted, 105–107
    OLS regression for linear bivariate model, 121–122
    polycentric models, 107–110
    sample SAS program for monocentric function fittings, 123–124

G

GAM, *see* Geographical analysis machine
Gamma distribution, 52
GARIN–LOWRY model, 221–225, 228
Gaussian elimination, 192, 221
Geary's C, 173, 174
Geocoding, 65
GeoDa, 53, 168, 183–185
Geodatabase model, 3
Geodetic distance, 20
Geographic approaches, rare event analysis
    analyzing rare events in small population, 149–150
    ISD and spatial-order methods, 150–152
    poisson-based regression analysis, 164–165

relationship between job access and homicide patterns, 155–163
scale-space clustering method, 152–155
Geographic coordinate system, 2
Geographical analysis machine (GAM), 168–169
Geographical Information System (GIS), 102–105, 110
Getis-Ord general G, 173, 177, 178, *see also* Geary's C
Getis's method, 188
G_i statistic, 173, 174, 178, 180, 188
GIS, *see* Geographic Information System
Global clustering, 168, 172–173, 178
Global empirical Bayes smoothing, 52
Global interpolation, 42, 45–46
Graphic user interface (GUI), 2
Gravity-based accessibility index, 82–83
Gravity-based methods
    density function approach to urban and regional structures, 98
    deriving urban density functions, 121
    GIS-based measures of spatial accessibility, 82–84
    measuring spatial accessibility to primary care physicians, 89–91, 94
    property for accessibility measures, 95
Gravity kernel, 59
Gravity method, 55, 57–63
Gravity potential, 60
Guangxi Autonomous Region, 38–42, *see also* China
GUI, *see* Graphic user interface

H

Haidian district, 144, *see also* China
Hamilton's model, 193, 213–214
Hanson model, 78
Harbin, 25, 68, 71, 72, *see also* China
Health care, 78, 82–83
Health care providers, 203–211
Health professional shortage areas (HPSAs), 36, 84
Health-related research, 167
Heilongjiang, 25, 69, 71, 72, *see also* China
Heterogeneity, 165
HIV/Aids, 150
Homicide rates, *see also* Illinois
    Chicago
        jobs access comparison at multiple levels, 155–163
        spatial regression analysis, 182–185
    use of geographic approaches in small populations, 149
Hot-spot analysis, 167, 178
Housing reform, 135

HPSAs, *see* Health professional shortage areas
Huff model
   defining fan bases, 66–68
   defining hinterlands of major cities in China,
      68, 69–71, 72
   gravity models for delineating trade areas,
      59–60, 62
      extensions, 61–62
   Reilly's law link and gravity models for
      delineating trade areas, 60–61
Hypothetical city, 225–230

I

IDW, *see* Inverse distance weighted method
Illinois, 175–180, *see also* Chicago
Illinois State Cancer Registry (ISCR), 175
ILP, *see* Integer linear programming
Impedance values, 226, 227
Inaccessibility, , 82
INFO files, 27–28, 204–205, 207
INFO format, 3
Input–output model, 231–232
Integer linear programming (ILP), 199
Integer programs, 199–202
Intercept, 122
Intermediate store sites, 56
Intersect, 10, *see also* Map overlays
Inverse distance weighted (IDW) method, 43, 46
ISCR, *see* Illinois State Cancer Registry
ISD method, 150–152

J

Jacobian matrix, 233, 234
Jilin, 25, 69, 72, *see also* China
Job accessibility, 78–79, 155–156, 182
Job locations, 196
Jobs, 155–163, *see also* Employment
Jobs–housing balance approach, 79
Join
   ArcGIS
      attribute and data management, 3–4
      mapping the population density pattern, 7
   Chicago
      analysis of urban density patterns, 115
      measuring spatial accessibility to primary
         care physicians, 87
   China
      analyzing Tai place-names by spatial
         smoothing using FCA, 39
      measuring travel distance between counties
         and major cities, 29, 30
      measuring wasteful commuting, 196

K

k nearest neighbors, 168
Kernal density, 41
Kernal estimation, 37–38, 41–42
Kriging, 44, 47

L

L matrix, *see* Valued-graph approach
Label-setting algorithm, 21–22
Lagrangian function, 74
Lambert Conformal Conic projection, 2
Land use intensity, 140, 142, 143, 144
Layer, *see* Source layer
Least-effect principle, 71
Liaoning, 25, 69, 72, *see also* China
LIMDEP statistical pattern, 165
Linear bivariate model, 121–122
Linear equations
   GARIN-LOWRY model, 221–225
   FORTRAN program for solving, 234–241
   input–output model, 231–232
   local interpolation method, 44
   simulating population/service employment
      distributions in hypothetical city,
      225–230
   solving a system, 219–221
      nonlinear equations, 232–234
Linear function, 101
Linear programming (LP)
   allocating health care providers, 203–211
   integer programming and location-allocation
      problems, 199–202
   simplex algorithm, 190–193
   wasteful commuting
      Hamilton's model, 213–214
      measuring, 193–199
      SAS program for the problem of
         measurement, 214–217
Linear regression, *see also* China; Illinois
   analysis of urban density patterns, 113, 114, 116
   function fittings by regression, 118–119
   social area analysis, 141
Links, 23
LISA, *see* Local indicator of spatial association
Local clusters, 168–169, 173–175, 178
Local indicator of spatial association (LISA), 173
Local interpolation, 42, 43–47
Local maxima, 153, 154, 161
Local minima, 153, 154, 161
Local Moran index, 173, 174, 178
Local polynomial interpolation, 43
Location, trade area analysis, 55

Location set covering problem (LSCP), 201, 202
Location-allocation problems, 199–202
Logarithmic functions, 101, 113
Logistic trend surface method, 43, 46
Log-transform, 101, 104, 106, 107
Lower triangle (matrix) structure, 220, 221
LP, *see* Linear programming
LSCP, *see* Location set covering problem
LU decomposition, 221, 228
LU factorization phase, 221
Lung cancer, 175, *see also* Cancer

M

Major League Baseball (MLB), 63–68
Manhattan distance, 19, 24–26, 28, 172
Map elements, 4
Map of scattered discrete points, 37
Map overlays, 10, 12, 85
Map projections, 2–3
Mapping
    China
        factor patterns and social area analysis,
            140
        Tai place-names by point-based spatial
            interpolation, 45–47
    Illinois
        cancer patterns and spatial cluster analysis,
            175–180
        job access versus homicide patterns at
            multiple levels, 158
        population density and analysis of urban
            density patterns, 111
    population density pattern, 4–8, 9
MATLAB language, 221
MAUP, *see* Modifiable areal unit problem
Maximum covering location problem (MCLP),
    201, 202
Maximum likelihood, 52
Maximum likelihood estimation, 184
Maximum likelihood estimator, 181, 182
Maximum likelihood test statistic, 169
MCI, *see* Multiplicative competitive interaction
    model
MCLP, *see* Maximum covering location problem
MCLP mandatory closeness constraints, 202
Melting process, 153, 155
Microsoft Excel, 103, 111, 113, 117
Mills–Muth economic model, 98, 120–121
MILP, *see* Mixed-integer programming problem
Minimum requirements approach, 222
Mixed-integer programming (MILP) problem, 199
MLB, *see* Major League Baseball

MNL, *see* Multinomial logit model
Modifiable areal unit problem (MAUP), 51–52,
    111, 119
Modified Gauss–Newton Method, 105–106
Modules, 2
Moments procedure, 52
Monocentric city, 104
Monocentric function fittings, 123–124
Monocentric model, *see also* Illinois
    analysis of urban density patterns, 116–117
        function fittings, 111–115, 118
    density function approach to urban and
        regional structures, 98, 99, 100
    function fittings by regressions, 101–105
    measuring spatial accessibility to primary care
        physicians, 86–87
Monte Carlo simulation approach, 169
Moran's I statistic, 172–173, 174, 177, 178
Moving averages, 35
Multichoice logistic model, 59
Multinomial logit (MNL) model, 62
Multinomial logit regression mode, 62
Multinuclei model, 135
Multiple ring buffer, 10, *see also* Map overlays
Multiple testing, 169
Multiplicative competitive interaction (MCI) model,
    62–63, 73
Multivariate regression model, 63
Multivariate statistical analysis methods
    cluster analysis, 131–134
    discriminant function analysis, 145–146
    principal components and factor analysis,
        127–131
    sample SAS program for factor and cluster
        analyses, 146–147
    social area analysis, 134–135
        Beijing, 135–143

N

NACJD, *see* National Archive of Criminal Justice
    Data
National Archive of Criminal Justice Data (NACJD),
    156
Nearest-neighbor method, 132–133
Negative exponential function, 98
Negative spatial autocorrelation, 174
Negative unit elasticity for housing demand, 120
Neighborhood dynamics, 140, 142, 143, 144
Neighborhoods, 49
Network, 23, 31
Network-based analysis, 207–211
Network coverage, 27

Network distance
    computing, 21–24
        simulations of population/employment
            distributions in hypothetical city,
            226–227
    measuring distances and time, 20
    measuring travel distance between counties and
        major cities in China, 29, 30
Network hierarchical weighting (NHW) method, 48
Network-overlaid algorithms, 48
Network time, 21–24
New Zealand, 78
Newling's model, 102, 104, 113, 114, 117
Newton's method, 232, 233
NHW, see Network hierarchical weighting method
NIMBY, see Not in My Backyard
Nodes, 23–24, 196
Nonbasic (service) economic activity, 221–222
Nonbasic sector, 222
Nonbasic variables, 192
Nongeographic strategies, 149
Nonlinear equations, 232–234
Nonlinear regression
    analysis of urban density patterns, 114
    function fittings, 105–107
        regression applied to density pattern analysis,
            117, 118–119
Nonrandom events, 167
Normalization, 224
Not in My Backyard (NIMBY), 175
Null hypothesis, 168

O

Objective function, 190, 194, 198
Object-oriented data model, 3
Oblique rotation, 130
O'Hare Airport, 115–116
Ohio, 31–32
OLS, see Ordinary least squares linear regression
One-step link, 31, 32
On-the-fly reprojections, 2
Optimal commute, 213, 214
Optional feasible point, 190
Ordinary least squares (OLS) linear regression
    function fittings
        monocentric models, 101
        nonlinear and weighted regressions, 105
    job access versus homicide patterns at multiple
        levels, 156, 160, 163
    linear bivariate model, 121–122
    regression implementations and GIS, 103
    spatial filtering methods for regression analysis,
        188

spatial regression, 181
    analysis of homicide patterns, 183–184,
        185, 186
Origin nodes, 27, 29, see also Nodes
Orthogonal rotation, 130
Overprediction, 122

P

PALINFO, 14
PCA, see Principal components analysis
PCFA, see Principal components factor model
Percentage errors, 105, 106
Physician–population ratio, 86
Physician services, 82–83
Physician shortage area, 79
Plane coordinate system, 2
p-Median problem, 200–201, 202
p-Median with a maximum distance constraint, 202
Point-based methods, 167, 169–170
Point-based spatial cluster analysis, 170, see also
        Cluster analysis
Point coverage, 40, 225, 226
Point distance, 115
Point Distance tool, 20
Poisson tests, 169
Poisson-based regression analysis, 150, 163–164
Polycentric models
    analysis of urban density patterns, 115–116
    density function approach to urban and
        regional structures, 100
    function fittings, 107–110
Polycentricity, 99
Polygon adjacency, 12–15
Polygon adjacency matrix, 14
Polygon-based analysis, 203–207
Polygon coverage
    allocating health care providers, 203
    Illinois
        analysis of urban density patterns, 111
        defining fan bases, 63, 65
        job access versus homicide patterns at
            multiple levels, 156
        measuring spatial accessibility to primary
            care physicians, 86
        spatial regression analysis of homicide
            patterns, 182
    regression implementations and GIS, 103
    simulations of population/employment
        distributions in hypothetical city, 225
Polygons, 153, 154
Population
    distributions and simulations in hypothetical
        city, 225–230

GARIN–LOWRY model in solving a system
    of linear equations, 221–223
Population density, 102, 111
Population density pattern, 4–8, 9
Population-weighted centroid, 85, *see also* Centroids
Potential, 60, 66, 70–71
Potential accessibility, 77
Potential model, 78
Power function, 101, 113, 114
Primary care physicians, 84–91
Principal components analysis (PCA)
    multivariate statistical analysis method, 127
    social area analysis, 138, 139
        use in data reduction, 127–131
Principal components factor analysis (PCFA), 128,
        130, 131, 137–183
PRJ, *see* Projection definition file
Probabilities, 66
Probability surface, 63, 66
Probability surface mapping, 66–68
Production–attraction constrained model, 60
Production-constrained model, 60
Projection, ArcGIS, 2
Projection definition file (PRJ), 2
Prostate cancer, 175–180
Proximal area, 69, 70
Proximal area method, 55, 56–57, 65–66

Q

Qinzhou Prefecture, 38–42, *see also* China
Quantitative analysis, 119
Queen contiguity, 14, 15, 16, 172, 183
Queries, 8

R

Radical basis functions, 44
Railroad distance, 69, 71
Railroads, 28
Randomization, 173, 174
Randomness of sample, 106
Raster data model, 3
Rational choice, 155
Ratios, 40
Reference points, 3
Reference zone, 143
Regionalization, 150
Regionalized empirical Bayes smoothing, 53
Regression, 102–105
Regression analysis, 55, 56, 158, 163, 187–188
Regression models, 42, 43, 140–141, 143
Reilly's law, 57–61

Relate, 7, 29, *see also* Join
Required commute, 214
Resident work locations, 196
Residential differentiation, 134–135
Residual sum squared (RSS), 105, 118, 122
Retail banks, 56
Revealed accessibility, 77
Road coverage, 207
Road network coverage, 225, 226, 227
Rook contiguity, 14, 15, 172
Root mean square, 42
Rotated factor patterns, 157
Rotation, 130–131
Rounded solution, 200
RSS, *see* Residual sum squared
Rural land use model, 99

S

Sales forecasting, 56
San Francisco, California, 174
SAR, *see* Simultaneous autoregressive model
SAS program
    analysis of urban density patterns, 111, 114–115
    cluster analysis, 134
        factor analysis, 146–147
    factor and principal components analysis, 129, 130
    function fittings
        monocentric, 123–124
        nonlinear and weighted regressions, 106–107
        polycentric, 110
    social area analysis, 137–138, 140
    solving linear programming problems, 193
    wasteful commuting
        measuring, 197–199
        problem and linear programming, 214–217
SaTScan, 168, 169, 170, 171
Scaled row pivoting, 221
Scale-space clustering method, 152–155, 158
Scale-space melting method, 155–163
Scale-space smoothing, 153
Scale-space theory, 152, 154
School districts, 48–51
Scree graph, 138
Scree graphy, 129, 130
Scree plot, 134
Sector model, 135
Sectoral structures, 143, 144
Sensitivity analysis, 40, 88, 91
Service (nonbasic) employment
    basic economic activity distinction of
        GARIN–LOWRY model, 221
    distribution and simulations in hypothetical city,
        225–230

Settlement areas, 38–42
SF1, *see* Summary file
Shapefile
  China
    point-based spatial cluster analysis Tai
      place-names, 170
    social area analysis, 137, 140
  Illinois
    analysis of urban density patterns, 111, 117
    defining fan bases, 63, 65
    job access versus homicide patterns at
      multiple levels, 158, 159, 161, 162, 163
    measuring spatial accessibility to primary
      care physicians, 84
    spatial regression analysis of homicide
      patterns, 183
  Ohio
    allocating health care providers, 203
    area-based data aggregation of
      neighborhoods and school districts, 49
    mapping the population density pattern, 5, 6
Shapefile model, 3
Sheffield method, 150
Shenyang, 25, 68, 71, 72, *see also* China
Shijingshan district, 144, *see also* China
Shortest-route problem, 21–23, 31–32
Simple join, 10, 11, *see* Join
Simple supply–demand ratio method, 78
Simplex algorithm, 190–193
Simultaneous autoregressive (SAR) model, 181
Single-linkage method, 132, 134
Slack variables, 191
Slope, 122
Small numbers problems, 35
Small population, 149–150, 158, 159
Social area analysis
  Beijing, 135–143
  factor analysis, 127–131
  multivariate statistical analysis, 134–135
  principal components factor model, 128, 120,
    131
  use of cluster analysis, 131
Social stress theory, 155
Socioeconomic status, 140, 142, 143, 144
Socioeconomic status factor, 134, 136
Socioeconomic variables, 157–158
Solution phase, 221
Source layer, 10
Space-filling curves, 152
Sparse format, 198, 199
Spatial accessibility, GIS-based measures and
    applications in health care
  issues, 77–79
    floating catchment area methods, 79–82
    gravity-based method, 82–84

primary care physician in the Chicago region,
    84–91
property for accessibility measures, 95–96
Spatial analysis tools, 8–12
Spatial autocorrelation, 167, 172, 173, 177
Spatial cluster analysis, 175–180, *see also* Cancer;
    Illinois
Spatial cluster analysis/spatial regression,
    applications on toponymical, cancer,
    and homicide studies
  area-based spatial cluster analysis, 172–175
  cancer patterns in Illinois, 175–180
  homicide patterns in Chicago, 182–185
  point-based, 168–169
  spatial filtering methods for regression analysis,
    187–188
  spatial regression, 181–182
  Tai place-names in China, 170–171
Spatial data, 4–8, 9
Spatial data management, 1–4
Spatial data models, 2–3
Spatial dependence, 181
Spatial error model, 181
Spatial error regression model, 184, 185, 186
Spatial filtering method, 187–188
Spatial homogeneity, 150
Spatial interaction model, 60
Spatial interpolation, 37, 42–45, 47–48
Spatial join, 9–10, 11, 12, 13
Spatial lag, 173
Spatial lag model, 181, 182, 185
Spatial moving average model, 181
Spatial proximity, 150
Spatial query, 9, 10, 12
Spatial query tool, 14
Spatial regression, 181–182
  analysis, 182–185
Spatial scan statistics, 169
Spatial smoothing, 52–53, 83, 91, 92
Spatial smoothing/interpolation
  aggregating data from census tracts to
    neighborhoods/school districts, 48–51
  area-based, 47–48
  China, Tai name-places
    analyzing, 38–42
    surface modeling and mapping, 45–47
  empirical Bayes estimation, 52–53
  point-based spatial interpolation, 42–45
  spatial smoothing, 35–38
Spatial structure, 140
Spatial weights, 172, 183
Spatially autoregressive model, 181, 182
Spatial-order method, 150–152
SPCS, *see* State Plane Coordinate System
Speed limit, 196

Splines, *see* Individual entries
Spread, 100, *see also* Decentralization
SQL, *see* Structured Query Language
Stand-alone table, 4
Standard form, linear programming, 190
Standardized coefficients, 129
State element date, 48
State Plane Coordinate System (SPCS), 2
Statistical software, 106
Strain theory, 155
Structure Query Language (SQL), 7, 8
Subdistricts, 136, 137, 143
Suburbanization, 99
Summarized join, 10, 11, *see also* Join
Summary file (SF1), 7
Supply/demand, spatial accessibility, 78
Supply-oriented accessibility measures, 78
Supply-to-demand area, 79
Supply-to-demand ratio, 80–81
Surface modeling, 45–47
Surface modeling techniques, 110T

T

2SFCA, *see* Two-step floating catchment area
        method
t Statistic, 143
TA, *see* Total accessibility
Tableau form, 191
Tables, 4, *see also* ArcGIS
Tai place-names, 38–42, 170–171, *see also* China
Tanner–Sherratt model, 102, 104, 113, 114, 117
Taylor expansion, 232, 233
TAZ, *see* Traffic analysis zone
Text file, 137, 138, 156
Thiessen polygons, 57, 58, 66
Thin-plate splines, 43, 44, 46
Threshold distance, 80, 81, 83, 86, 88
Threshold population, 150
Threshold time, 91, 93
Threshold value, 151
TIGER conversion tool, *see* Conversion tool
TIGER files, *see* Topologically Integrated
        Geographical Encoding and
        Referencing files
TIGER/Line files, 23
Tobler's first law of geography, 42, 43
Topologically Integrated Geographical Encoding
        and Referencing (TIGER) files, 5, 36,
        48, 194, 195
Topology builders, 23
Townships, 116–117, 118
Tract perimeters, 226

Trade area analysis, GIS-based
    basic methods, 56–57
    defining fan based, 63–68
    defining hinterlands of major cities in China,
        68–71
    economic foundations of gravity model,
        73–75
    gravity-based methods for delineating, 57–63
Traffic analysis zone (TAZ)
    area-based spatial interpolation, 48
    measuring spatial accessibility to primary care
        physicians, 36
    measuring wasteful commuting, 194, 195,
        198, 199
Transportation, improvements, 98, 99
Transportation network, 23–24, 221, 230
Travel distance
    measuring between counties and major cities
        in China, 26–31
    simulating distributions of populations/service
        employment in hypothetical city,
        227, 230
Travel friction coefficient, 89, 229–230
Travel impedance, 203
Travel speed, 31, 89, 226
Travel time, 81, 85, 89, 90, 226
TREE procedure, 134
Trend line, 118, 119
Trend line tool, 113
Trend surface analysis, 42, 43, 45
Two-step floating catchment area (2SFCA) method
    GIS-based measures of spatial accessibility,
        80–82
        earlier methods, 79–80
        gravity-based method comparison, 83–84
    measuring spatial accessibility to primary care
        physicians, 85–89, 90, 92, 93property
        for accessibility measures, 95
        Total accessibility (TA), 95

U

Underprediction, 122
Unemployment, 155–156, *see also* Employment;
        Illinois
Union, 10, *see also* Map overlays
Unique factors, 128
Universal Transverse Mercator (UTM), 2, 5, 49
Upper triangle (matrix) structure, 220, 221
Urban area, 222, 224, *see also* Linear equations
Urban density functions, 120–121
Urban density patterns, 110–117
Urban element data, 48

Urban hinterlands, 68–71, 72, *see also* China
Urban land use model, *see* Alonso's urban land use
    model
Urban land use reform, 135
Urban polarization, 99
Urban structure, 135
Urban system planning, 68
Urban/regional structures, 97–101
U.S. Cellular Field, 64
Utility function, 73
Utility maximization, 121
UTM, *see* Universal Transverse Mercator

V

Validation techniques, 42
Valued-graph approach, 31–32
Varimax rotation, 130, 139
VBA, *see* Visual Basic statement
Vector data model, 3
Visual Basic (VBA) statement, 6

W

Ward's method, 133, 134
Wasteful commuting, 193–199, *see also* Ohio
Weighted averages, 158–159, 162
Weighted regression, 105–107, 114, 118, 119
White Sox, *see* Chicago White Sox
White's model, 193–194
Wrigley Field, 64

X

Xuanwu district, 144, *see also* China

Z

z Test, 177, 178
0-1 (binary) programming problem, 200
Zip code areas centroid, 85, *see also* Centroids
Zonal structures, 143, 144

Related Titles

GIS and Geocomputation
Peter Atkinson
ISBN: 0-748-40928-9

Programming ArcObjects with VBA Task Oriented Approach
Kang-Tsung Chang
ISBN: 0-849-32781-4

Introduction to Mathematical Techniques Used in GIS
Peter Dale
ISBN: 0-415-33414-4